AF360915

Smart Energy, Plasma and Nuclear Systems

Smart Energy, Plasma and Nuclear Systems

Editor

Hossam A. Gabbar

MDPI • Basel • Beijing • Wuhan • Barcelona • Belgrade • Manchester • Tokyo • Cluj • Tianjin

Editor
Hossam A. Gabbar
University of Ontario Institute of
Technology
Canada

Editorial Office
MDPI
St. Alban-Anlage 66
4052 Basel, Switzerland

This is a reprint of articles from the Special Issue published online in the open access journal *Energies* (ISSN 1996-1073) (available at: https://www.mdpi.com/journal/energies/special_issues/SEGE_SPANS2019).

For citation purposes, cite each article independently as indicated on the article page online and as indicated below:

LastName, A.A.; LastName, B.B.; LastName, C.C. Article Title. *Journal Name* **Year**, *Volume Number*, Page Range.

ISBN 978-3-0365-0752-1 (Hbk)
ISBN 978-3-0365-0753-8 (PDF)

Contents

About the Editor

Hossam A. Gabbar (Dr.) Dr. Gabbar is a full Professor in the University of Ontario Institute of Technology (UOIT) in the Faculty of Energy Systems and Nuclear Science, and is cross appointed in the Faculty of Engineering and Applied Science, where he has established both the Energy Safety and Control Lab (ESCL) and Advanced Plasma Engineering Lab. He is leading national and international research in the areas of nuclear safety and control, advanced plasma systems and their application in nuclear fusion, as well as in the control and protection of smart grids and micro energy grids. He is leading research efforts in Canada, achieving international recognition for his work in the field of energy safety and control for nuclear and energy production facilities.

Preface to "Smart Energy, Plasma and Nuclear Systems"

This Special Issue on the topics of smart energy, plasma and nuclear systems contains a collection of five extended papers from the SEGE (International Conference on Smart Energy Grid Engineering) 2019 and SPAN (Symposium on Plasma And Nuclear Systems) 2019. The SEGE conference aims to provide an opportunity to discuss various engineering challenges of smart energy grid design and operation by focusing on advanced methods and practices for designing different components and integrating these within the grid. It also provides a forum for researchers from academia and professionals from industry—in addition to government regulators—to tackle these challenges and discuss and exchange knowledge and best practices about the design and implementation of smart energy grids. The Symposium on Plasma and Nuclear Systems (SPANS) is connected with SEGE. SPANS provides a forum for researchers from academia and industry to present and discuss the latest research innovations in nuclear and plasma systems. SPANS will provide attendees with state-of-the-art research and technologies, while engaing in active discussions with industry. Additionally, it will provide industry with opportunities to promote their products and business cases. Attendees from regulators and standards will engage in fruitful discussions on how R&D is linked with regulations and standards. The 2019 edition of SEGE was held in Oshawa, Canada, and attracted a total of 112 regular paper submissions, spanning numerous active and emerging topic areas. The conference program committee selected 60 papers to be presented at the conference and published in the conference proceedings. The five extended papers for this special issue were selected from among all the accepted papers by the Special Issue Guest Editor Dr. Hossam A. Babbar, based on the relevance to the journal and the reviews of the conference version of the papers. The authors were asked to revise the conference paper for journal publication and in accordance with the customary practice of adding 30% new material. The revised papers, again, went through the normal journal-style review process and are now finally presented to readers in this Special Issue. We greatly appreciate the willingness of the authors in helping to organize this Special Issue.

Hossam A. Gabbar
Editor

Article

Energy Analysis for the Connection of the Nuclear Reactor DEMO to the European Electrical Grid

Sergio Ciattaglia [1], Maria Carmen Falvo [2,*], Alessandro Lampasi [3] and Matteo Proietti Cosimi [2]

[1] EUROfusion Consortium, 85748 Garching, Germany; sergio.ciattaglia@euro-fusion.org
[2] DIAEE—Department of Astronautics, Energy and Electrical Engineering, University of Rome Sapienza, 00184 Rome, Italy; m.proietticosimi@gmail.com
[3] ENEA Frascati, 00044 Frascati, Rome, Italy; alessandro.lampasi@enea.it
* Correspondence: mariacarmen.falvo@uniroma1.it

Received: 31 March 2020; Accepted: 22 April 2020; Published: 1 May 2020

Abstract: Towards the middle of the current century, the DEMOnstration power plant, DEMO, will start operating as the first nuclear fusion reactor capable of supplying its own loads and of providing electrical power to the European electrical grid. The presence of such a unique and peculiar facility in the European transmission system involves many issues that have to be faced in the project phase. This work represents the first study linking the operation of the nuclear fusion power plant DEMO to the actual requirements for its correct functioning as a facility connected to the power systems. In order to build this link, the present work reports the analysis of the requirements that this unconventional power-generating facility should fulfill for the proper connection and operation in the European electrical grid. Through this analysis, the study reaches its main objectives, which are the definition of the limitations of the current design choices in terms of power-generating capability and the preliminary evaluation of advantages and disadvantages that the possible configurations for the connection of the facility to the European electrical grid can have. In reference to the second objective, the work makes possible a first attempt at defining the features of the point of connection to the European grid, whose knowledge will be useful in the future, for the choice of the real construction site.

Keywords: nuclear fusion; tokamak; generation power plant; power system; electrical transmission grid

---

1. Introduction

The European roadmap to fusion energy, summarized in Figure 1, includes the DEMOnstration power plant (generally identified as DEMO), which represents the first fusion reactor designed to supply electrical power to the electrical grid to which it will be connected [1]. With the start of its operation, currently set for the middle of the 21st century, DEMO could be a revolution in the world of nuclear fusion power and in general in the world of power generation.

The first and fundamental step of the roadmap however is ITER, a research tokamak project currently under construction in France that is foreseen to be operative in around five years. ITER is not designed to generate electrical power, but it is designed to achieve five goals that are essential for the continuation of the research in this field [2]:

- It will generate 500 MW of fusion thermal power during a relatively prolonged fusion time of 400 s;
- It will demonstrate the effectiveness of new technologies for heating, control, diagnostics, cryogenics and remote maintenance, applied to fusion reactors;
- It will achieve a deuterium-tritium plasma capable of self-sustainment;
- It will test the breeding blanket technology and so the possibility of producing the tritium needed for the fusion reactions inside the reactor itself;

- It will demonstrate the safety characteristics of a fusion device.

DEMO will largely build on the ITER experience; indeed, its construction will start after several years of ITER operation. Now, the design of DEMO is based on five main objectives [3]:

- Conversion of fusion thermal power into electricity for several hundreds of megawatts;
- Achievement of tritium self-sufficiency (the tritium produced through the breeding blanket technology is higher than that consumed during the fusion reaction);
- Reasonable availability of up to several full-power years;
- Minimization of radioactive wastes, with no-long-term storage;
- Extrapolation to a commercial fusion power plant.

Considering the achievements that both the facilities should reach, the actual characteristic that distinguishes DEMO from ITER is the size and consequently the possibility of achieving higher values of fusion gain factor and longer fusion time. The fusion gain factor is defined as the thermal power produced inside the reactor during the fusion reaction divided by the thermal power delivered to the plasma during the operation. In particular, DEMO being bigger than its predecessor, it is designed to reach a fusion gain factor between 10 and 50 [1], while this value for ITER is foreseen to be around 10 [1]. One of the aims of current studies on this topic is to understand if this gain factor is high enough to allow a feasible supply of electrical energy to the grid, also because the electrical power systems of tokamaks like ITER or DEMO are larger and more complex than those of nuclear-fission power plants.

As for every tokamak, the main limitation for DEMO is the impossibility of maintaining the fusion reactions for an indefinite time [4]. This is an intrinsic characteristic of tokamaks, and it is related to the need of charging and discharging the central solenoid (CS) system that generates and confines the plasma current. This essentially means that the operation of DEMO is variable, and in particular, it is divided in several phases that will be presented in following section.

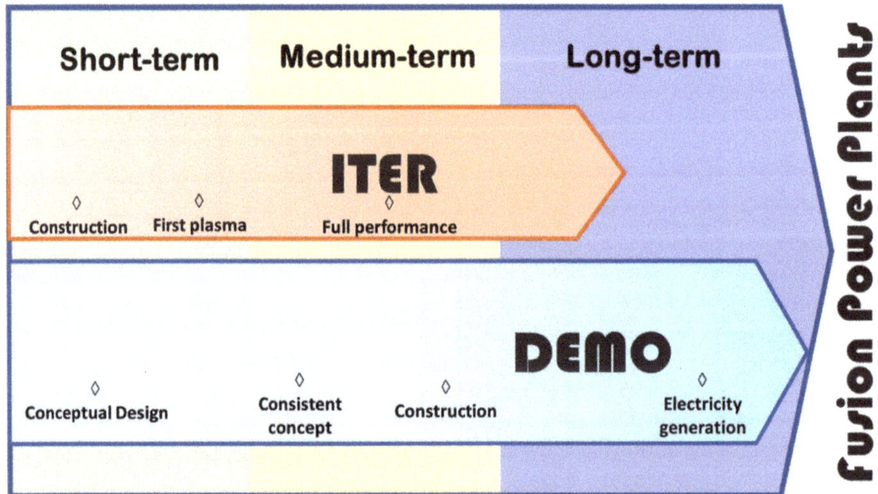

Figure 1. Summary of the EUROfusion Roadmap to fusion energy.

Now, the researchers are trying to understand how and how much the non-generation time can be reduced, even if the generation time is already foreseen to be at least one order of magnitude greater than the non-generation time. From the structural point of view, the need to optimize this characteristic led to the identification of two possible alternative configurations. The first one involves the direct coupling between the Primary Heat Transfer System (PHTS, the system that extracts the thermal power from the reactor walls) and the Power Conversion System (PCS), which reflects the variability of the thermal output into the electrical power output. The second one involves the indirect coupling between the PHTS and the PCS, which allows to decouple the variable thermal output from the electrical power output, with the interposition of an Intermediate Heat Transfer System (IHTS) with an Energy Storage System (ESS) based on molten salts [3], allowing a constant electrical power output at the generator level.

The article includes five sections. Section 2 defines the main characteristics of DEMO, and it introduces its operational phases. Section 3 focuses on the electrical generator, defining the limitations of the possible coupling configurations, through the analysis of the European Network of Transmission System Operators (ENTSO-E) requirements for generators. Section 4 presents three possible connection solutions for the DEMO facility, considering both the demand and the generation, providing an overview of the advantages and disadvantages of each solution. In Section 5, starting from the input and output power profiles, some features of the point of connection to the grid are evaluated. Section 6 resumes the conclusions of the study.

2. DEMO Features and Operational Phases

DEMO is foreseen to generate a fusion thermal power inside its reactor that has been evaluated to be in the order of 2 GWth [5]. Now, two solutions are under study for the thermal power extraction and accordingly for the PHTS. The first solution exploits a mature technology, which is the water cooling, also used in fission nuclear plants for its simplicity and reliability. The second solution instead foresees the use of helium for the PHTS, which seems to be promising for future applications but is still a relatively new technology. Conventionally, the first solution is identified as Water Cooled Lithium Lead (WCLL), where lithium lead refers to the technology adopted inside the reactor wall, and the second one as Helium Cooled Pebble Bed (HCPB) [6].

In case of direct coupling between the PHTS and the PCS, both the WCLL and the HCPB configurations provide for the implementation of a Rankine cycle for the thermal power conversion. This means that in case of HCPB, the helium cools the reactor (PHTS), and then, it exchanges the extracted power with the cycle working fluid, namely water. Moreover, in case of indirect coupling, the working cycle is a Rankine cycle, but in this case, the decoupling between the PCS and the PHTS makes the overall operation independent of the type of fluid circulating in the PHTS. From now on, more efforts are being focused on the WCLL configuration and so more data are available. This study will mostly refer to this solution. In any case, several results and procedures can be applied both to the WCLL and to the HCPB configurations.

For the limits of the tokamak technology, the thermal power is generated only during a portion of the operation time, which is generally identified as the "Burn Flat-Top" phase. A smaller power can be generated in the other phases for thermal inertia of the materials, nuclear reactions and other phenomena but always, as a consequence of the Burn Flat-Top. In the present configuration, this phase should last around 7200 s (2 h). Between each burn phase, other phases are conventionally identified during which no relevant thermal power is extracted. These phases are resumed in Table 1, with their respective duration.

For completeness, a brief description of the phases is reported. During CS pre-magnetization, the CS, that is, the core of the magnet system, is energized in order to be able to generate a power pulse strong enough to generate the plasma ignition that is represented by the Breakdown phase. Once the plasma has been generated inside the reactor, it must be heated up to the temperatures needed to have a sustainable rate of fusion reactions, and this is done during the Plasma Ramp-Up and Heating Flat-Top phases, mainly using the additional heating (AH) systems. Now, the most likely solution allows for the use of three technologies for the AH: Electron Cyclotron Resonance Heating (ECRH), Ion Cyclotron Resonance Heating (ICRH) and Neutral Beam Injection (NBI) [7]. When the temperature reaches the order of 10^7 K, the conditions inside the reactor allow to have self-sustained fusion reactions happening, and this identifies the Burn Flat-Top phase. During this phase, the power demand from the magnet system and the AH is minimum, while the thermal power generated is maximum. At the end of the burn phase, the reactor has to be brought back to the initial conditions avoiding shocks in the plasma, and this is done during the Plasma Ramp-Down and the Dwell time.

Table 1. Plasma phases.

Phase Initial Time	Phase Final Time	Phase Duration	Phase Name
−500 s	0 s	500 s	CS pre-magnetization
0 s	1.4 s	1.4 s	Plasma Breakdown
1.4 s	184 s	≈183 s	Plasma Ramp-Up
184 s	194 s	10 s	Heating Flat-Top
194 s	7394 s	7200 s	Burn Flat-Top
7394 s	7540 s	≈146 s	Plasma Ramp-Down
7540 s	7740 s	200 s	Dwell time

The whole operation of DEMO, in terms of input and output power of the facility, can be described referring to the phases resumed in Table 1.

For what concerns the power needed to operate the facility, two types of loads have to be considered: the steady-state loads and the pulsed loads. The first ones require a steady-state 50 Hz voltage input, i.e., auxiliaries, cryogenics, pumps or compressors for the PHTS, etc. The pulsed loads, i.e., magnet system and AH devices, have to be supplied with variable voltage waveforms, and so, they will be provided with complex power conversion systems that are currently under study. The steady state and pulsed loads are supplied by dedicated substations and distribution systems [5], respectively, the Steady State Electrical Power System (SS EPS) and the Pulsed Power Electrical Power System (PP EPS). By convention, the PP EPS is identified with its two subsystems, the Coil Power Supply Pulsed EPS (CPSP EPS) and the Heating Power Supply Pulsed EPS (HPSP EPS), which supply the magnets system and the AH devices, respectively. From the generation point of view, instead, the generator subsystem is here conventionally referred to as Electrical Generator EPS (EG EPS). The connection node and the High Voltage/Medium Voltage (HV/MV) transformation sub-station are defined as High Voltage Switchyard EPS (HVS EPS).

In order to have a preliminary overview of the facility's demand, Figure 2 presents a qualitative profile of the input active power, derived from the data available in the EUROfusion private database.

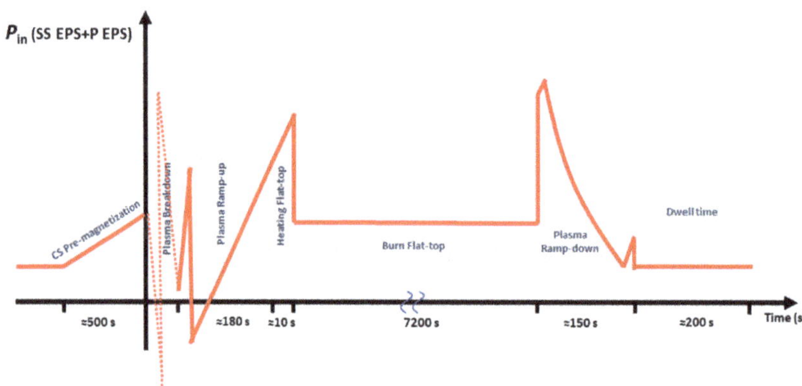

Figure 2. Qualitative profile of DEMOnstration power plant (DEMO) active power demand during one cycle.

Regarding the power generation, of course, the direct and indirect coupling cases have to be separately considered to evaluate the output power profiles.

Despite being the simplest solution form the constructive point of view, the direct coupling between the PHTS and the PCS leads to several thermomechanical and electrical concerns. Regarding the thermomechanical aspects, several studies, not public but available for the authors as researchers involved in DEMO project on the Eurofusion database, assess the impossibility of operating the turbine with a completely direct coupling with the reactor. Firstly, the thermomechanical stresses due to the abrupt changes in the steam mass flow rate would be unsustainable for the turbine, leading to premature failures of the turbine itself. To limit these cyclical stresses, the maximum steam mass flow rate has to be reached with a ramp. In particular, the nominal power of the turbine has to be reached with an increase of 10% of the nominal power per minute. Another limit concerns the minimum power at which the turbine can be operated, that is, the minimum steam flow rate that can be supplied. In this case, the problem affects both the thermomechanical and the electrical aspects. Indeed, from the thermomechanical point of view, the turbine would suffer from the cyclical start and stop procedures, while from the electrical point of view the generator cannot lose the synchronism with the grid, so it has to be kept spinning. This means that the turbine has to be supplied with the proper mass flow rate of steam also during the non-generation time. In particular, the minimum power of the turbine has been set to 10% of the nominal power. During the reactor non-generation time, the solution currently under study allows for the implementation of a small electrically heated molten-salts loop, designed to provide the proper mass flow rate of steam to the turbine.

Referring to the latest studies, not public but available for the authors as researchers involved in DEMO project on the Eurofusion database, the nominal power of the steam turbine has been evaluated to be around 790 MW. Considering a 5% value for the losses due to the coupling between the turbine and the synchronous generator, the nominal value for the electrical power output can be estimated to be about 750 MW. Considering this nominal value and the limits previously reported in case of direct coupling between the PHTS and the PCS, the resulting active power profile is shown in Figure 3.

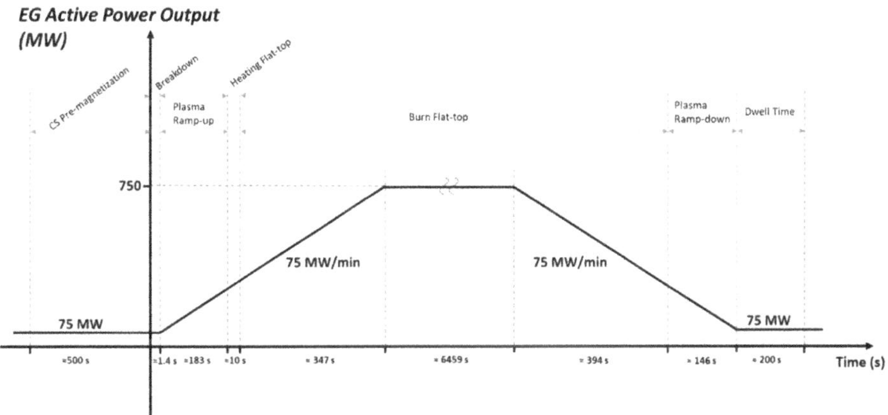

Figure 3. Electrical Generator (EG) active power output profile in case of direct coupling between the Primary Heat Transfer System (PHTS) and the Power Conversion System (PCS).

For the indirect coupling configuration, the situation is more complex from the constructive point of view, due to the IHTS and the molten-salts ESS, but the management of the turbine and of the generator is simpler. Indeed, in this case the steam flow rate supplied to the turbine can be maintained practically constant during the whole operation of the facility, like in a conventional power plant, minimizing both the thermomechanical and the electrical stresses.

Referring to the latest studies available on the EUROfusion private database, the nominal power of the turbine in case of indirect coupling configuration is around 675 MW. Considering the same value used in the previous case for the efficiency of the turbine-synchronous generator coupling (0.95),

the resulting active power output at the EG level is around 640 MW. The resulting profile is reported in
Figure 4.

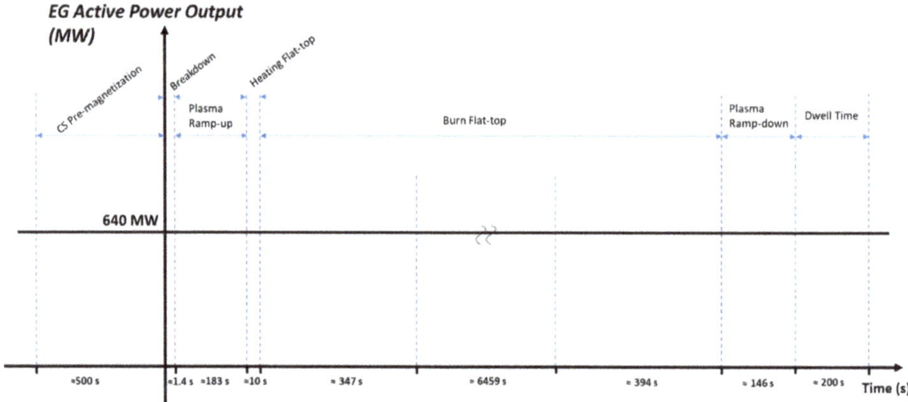

Figure 4. EG active power output profile in case of indirect coupling between the PHTS and the PCS.

3. Connection of DEMO EG to the European Electrical Grid

Being the first fusion reactor able to deliver electrical power to the external grid, DEMO has to
face new issues with respect to its predecessors. Of course, as it has been mentioned before, DEMO is
not a conventional power plant. However, since it uses a conventional thermodynamic cycle for the
conversion and is connected to the grid through a synchronous generator, now it has to be considered
as a conventional power plant from the regulatory point of view.

In reference to the current European power systems regulation, as a synchronous generator
connected to the transmission grid, for a proper operation, DEMO has to fulfil the requirements set by
the ENTSO-E. Specifically, these requirements are defined in the ENTSO-E Regulation of 14 April 2016
"Establishing a network code on requirements for grid connection of generators" (commonly said RfG
network code) [8].

In what concerns synchronous generators, which in the code are referred to as synchronous
Power Generating Modules (PGMs), the RfG distinguishes four categories, according to the voltage
level of the connection and to power rating of the facility (identified as maximum capacity inside the
regulation). The categories reported in the RfG are the following:

- Type A PGM, characterized by a voltage level at the connection point below 110 kV and maximum
  capacity of 0.8 kW or more;
- Type B PGM, characterized by voltage level at the connection point below 110 kV and maximum
  capacity between 1 MW and 50 MW;
- Type C PGM, characterized by voltage level at the connection point below 110 kV and maximum
  capacity between 50 MW and 75 MW;
- Type D PGM, characterized by any voltage level at the connection point and maximum capacity
  higher than 75 MW.

Therefore, for the specific case of DEMO, the requirements for Type D synchronous PGMs,
namely those with a power rating higher than 75 MW, have to be considered. The requirements for
Type D PGMs deal with:

1. Frequency stability;
2. Limited frequency sensitive mode—over-frequency;
3. Admissible power reduction;

4. Limited frequency sensitive mode—under-frequency;
5. Frequency sensitive mode;
6. Capability of maintaining constant output;
7. System restoration;
8. Voltage stability;
9. Robustness;
10. General system management;
11. Specific requirements for synchronous PGMs.

Since the research is still in a preliminary design phase, there is not enough information to analyze all the aspects of the RfG. Therefore, for the purpose of this study, we will refer only to the main aspects, i.e., those that can have an actual impact on the current design choices. In particular, we will consider the items in the list above from 1 to 7 and 11. The items from 1 to 6 deal with power generation and control, while items 7 and 11, respectively, deal with the capability of restoring the system after a shutdown of the grid and with the reactive power capability of the synchronous generator.

3.1. Constant Power Output Requirement

The frequency stability requirement defines the frequency ranges, and respective time intervals, for which the PGM shall be able to remain connected and operate in the grid. Moreover, for item 6 in the list above, in this range, the PGM shall be capable of maintaining a constant output at its target active power value. While in case of indirect coupling this requirement is fulfilled (Figure 4), it represents the strongest practical limitation to the choice of a direct coupling configuration, due to the variability of the EG power output. Referring to the profile of the active power output reported in Figure 3, considering the large variation of the output power (from 75 to 750 MW) and the fast variation in time, no compensation could be possible on the electrical side. Therefore, if the direct coupling configuration has to be adopted, the only way to fulfill the constant output requirement is to act on the mechanical power and so on the thermal power provided to the turbine or to act outside the main turbine-generator (TG) group. Therefore, the constant output power can be achieved in two main ways: One way is to make the TG work at the same active power set-point during all the phases; the other way is to couple the TG with an auxiliary generation set, even based on a totally different technology.

The first way can be seen as a category of solutions, all based on feeding a constant mass flow rate of steam to the steam turbine. Feeding always the same mass flow rate to the steam turbine essentially means to have an auxiliary steam generator. Actually, in the direct coupling PHTS-PCS scheme, one auxiliary steam generator is already foreseen, but it only provides the steam needed to operate the TG at the 10% of its nominal power (as mentioned in the previous section). Coherently with the volumes involved, the ESS loop could be foreseen with a higher rating, in order to guarantee up to 100% of the nominal mass flow rate of steam to the TG. Since the ESS loop is currently foreseen to be fed by an electric heater, it has to be considered that the net output power of the facility will be lower. Moreover, the convenience of the direct coupling approach, which is mainly the simplicity of construction with respect to the indirect coupling one, could be no longer that evident. Nonetheless, this solution could be even more complex than the pure indirect cycle.

The second solution is even more complex and expensive with respect to the first one, since it requires the implementation of a secondary generation set able to compensate the generation during the non-production time of the reactor. Since the power rating of the auxiliary generation set should be comparable with that of the first one, this kind of solution appears to be unlikely to be implemented. At the connection with the external grid, the sum of the two generation sets would appear as a good approximation of a conventional Type D PGM, but actually, it would have an installed maximum power capacity which is around two times the nominal one.

3.2. Power Control Capability Requirements

The Limited Frequency Sensitive Mode—Over/Under frequency (LFSM-O/U) is an operation policy, activated by the Transmission System Operator (TSO) when the grid is an emergency state of over/under frequency and needs a fast decrease/increase of active power generation [9]. The Frequency Sensitive Mode (FSM) instead represents the ordinary operating mode of a PGM, in which the power output changes in response to a change in the frequency of the system, in such a way that it supports the recovery of the target frequency [10]. Therefore, all the points from 2 to 5 deal with the power control capability of the PGM. In particular, the most restrictive value defined in the RfG is 10% (so $\Delta P_{max} = 0.1\ P_{max}$) power control capability, which is provided for the FSM operating mode. It is important to stress out that the provision of the power control capability is limited by the minimum regulating level and by the maximum capacity of the PGM. In addition, the ambient conditions and the limitations on the operation near maximum capacity at low frequency of the PGM have to be considered.

Regarding the indirect cycle configuration, while the stable and continuous operation with a fixed active power set point should be guaranteed thanks to the IHTS, evaluations on the molten salts ESS-water coupling should be performed, in order to define the potential of the system in terms of rate of change of active power.

In case of direct coupling, moving from the profile in Figure 3, the continuous and dashed red lines in Figure 5 represent the profile that the generation should be able to maintain in order to operate in the European Network as a conventional Type D synchronous PGM. In particular, the continuous line represents the maximum capacity operation that by definition the PGM should be capable of providing continuously during the ordinary operation in the grid. The dashed line instead defines the active power lower limit for the operation in case of FSM, with the power variation set to the maximum value of 10% of the maximum capacity as a conservative solution, since its actual value depends on the agreement carried out between the facility owner and the relevant TSO.

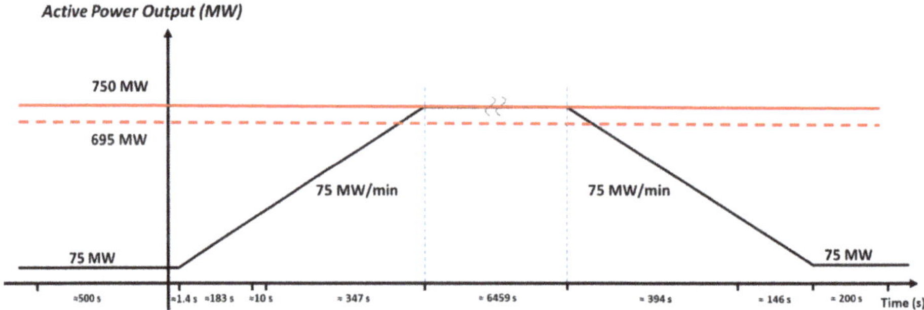

Figure 5. Active power output adapted to the European Network of Transmission System Operators (ENTSO-E) requirements.

Concerning the red dashed line, the capability of lowering the active power output, since the fusion thermal power has to be extracted in any case from the reactor (referring to generation time), it is sufficient to use suitable bypass valves upstream from the steam turbine. Of course, the solution must be compliant with the capacity of the condenser downstream the turbine that should be able to elaborate additional flows of steam and not only the double-phase flow coming from the turbine itself. For the direct coupling configuration, this condition should be already fulfilled since during the power ramps a variable portion of steam bypasses the steam turbine, so there could be no need for adjustments in these terms.

3.3. System Restoration Requirements

Concerning the system restoration, the relevant TSO may require the black start capability. A PGM with the black start capability shall be able to restart from a shutdown without the external grid electrical energy supply. This kind of service may be possible in case of indirect coupling, depending on the size of the reserve and on the duration of the shutdown, but it is not compliant with the operation in case of direct coupling. Indeed, in case of a network shutdown, the reactor is likely to be shut down in turn, since its operation and security strongly depend on the network supply. Moreover, for the same reason, the reactor cannot be restarted without the external grid, leading to the impossibility of a black start of the PGM.

3.4. Specific Requirements for Type D Synchronous PGMs

The operation of a Type D synchronous PGM requires additional specifications, mainly in terms of voltage stability and so about reactive power capability at and below the maximum capacity. From these requirements, we can extrapolate the maximum value for the Q/P_{max} ratio that a Type D synchronous PGM connected to the European electrical grid should be able to provide. This means that we can evaluate a preliminary power rating, in terms of apparent power, of the synchronous machine that has to be coupled with the steam turbine. Since the maximum value set in the RfG for this ratio is 0.65, which means $\cos\phi = 0.84$ (a common value for machines of this rating that can be found also in literature [11]), we will have that:

- In case of direct coupling, the maximum active power is equal to 750 MW during the burn phase; this means that the generator has to be able to supply $Q_{max\_direct} = 0.65 \cdot P_{max} \approx 490$ MVAr, and this results in an apparent power rating of the machine of around 900 MVA;
- In case of indirect coupling, the maximum active power is equal to 640 MW; this means that the generator has to be able to supply $Q_{max\_indirect} = 0.65 \cdot P_{max} \approx 416$ MVAr, and this results in an apparent power rating of the machine of around 765 MVA.

4. Layout Options for HV Switchyard Including the Generator

The results reported in this section take as reference the regulation on the connection of new facilities provided by the Italian TSO, Terna [12], considering that the Italian policies are based on the ENTSO-E prescriptions, so this approach is more conservative with respect to an approach based on the European requirements.

Relying on the guidelines provided by Terna, the connection of new facilities to the national grid must be planned in accordance with a proper procedure that is outlined in Figure 6.

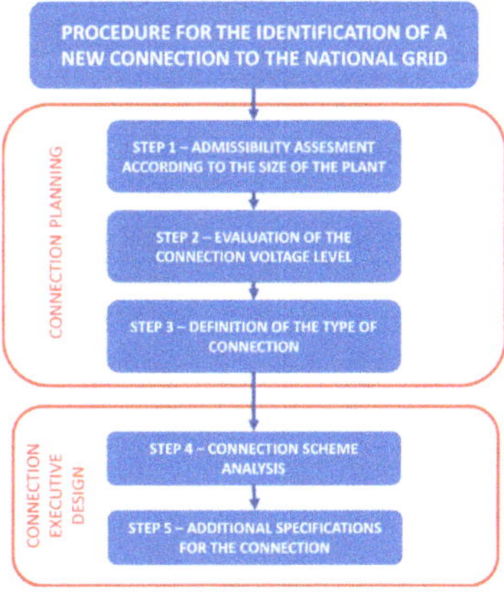

Figure 6. Procedure for the connection of generators and consumers to the European grid.

In particular, considering that the project is still in a preliminary design phase, the first three steps reported in Figure 6 will be considered in the analysis and so those that deal with the connection planning and not with the executive design, which are:

- The definition of the minimum value of power rating of the facility for which the connection has to be implemented in the transmission network, which is 10 MW; in particular, in case of both generation and demand facilities, the value to consider is the highest between the maximum injection and the maximum demand.
- The definition of the voltage level of the point of connection of the new facilities.
- The definition of the type of scheme to be implemented for the connection of the new facilities.

For the definition of the voltage level and of the connection scheme, the technical document [12] provides a table summarizing standard solutions, partially reported in Table 2.

Table 2. Standard solutions for voltage level and connection scheme of new generation or demand facilities.

	Power Rating of the Facility	Nominal Voltage Level	Standard Solutions	
			Radial	In-and-Out
Generation	100–250 MW	120–150 kV	Yes	No
	200–350 MW	220–380 kV	Yes	Single busbar + bypass
	>350 MW	380 kV	Yes	Double bus bar
Demand	20–50 MW	120–150 kV	Yes	Single busbar
	30–100 MW	120–150 kV	Yes	Single busbar
	>100 MW	220–380 kV	Yes	Single busbar + bypass

The in-and-out connection is performed with the implementation of a new station on an existing transmission line. This means that the new station is supplied by two different transmission lines, coming from two different nodes. Therefore, the facility can ideally operate even when one of the lines is out of service. The radial connection is similar to the in-and-out one, but the connection starts from an existing station of the transmission line. This solution is generally adopted when the distance between the existing station and the facility is lower than 10 km.

Keeping in mind this simplified procedure to evaluate the connection characteristics, three possible configurations have been considered for the connection of DEMO to the European transmission grid:

- Single Point of Delivery (POD); EG EPS, SS EPS and PP EPS are connected to the same HV Switchyard;
- Double POD with EG-dedicated node; SS EPS and PP EPS are connected to one HV Switchyard, while the EG EPS is connected to another HV Switchyard;
- Double POD with PP EPS-dedicated node; SS EPS and EG EPS are connected to one HV Switchyard, while the PP EPS is connected to another HV Switchyard.

In the following subsections these solutions will be analyzed starting from power profiles elaborated in the DIgSILENT PowerFactory simulation environment [13], moving from the data available on the EUROfusion database in terms of production and consumption of the facility. Concerning the demand, the data refers to the assisted breakdown scenario, which allows for the use of EHCR during the breakdown phase to assist the magnets system, lowering the power peak required for the plasma ignition.

4.1. Single POD

The single POD configuration is the one that requires the simplest implementation. In case of single POD, DEMO would fall in the category of power generating/demanding facilities. This means that the value of power to be considered in the evaluation of the connection scheme is the highest between the generation maximum power and the demand maximum power [12].

In Figure 7, the preliminary active power profiles are presented, in case of direct and indirect coupling.

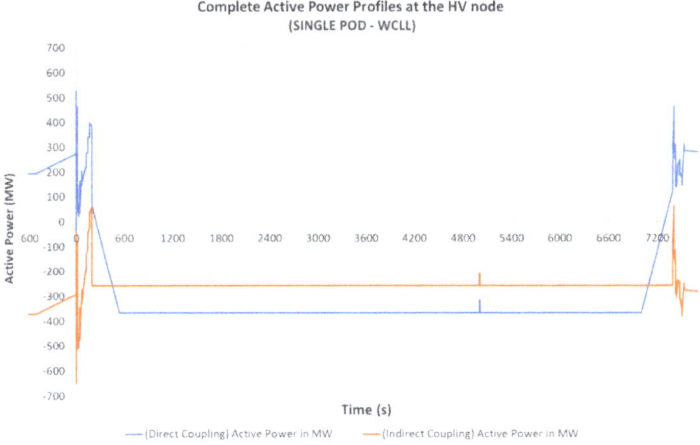

Figure 7. Power profiles at the HV node in case of single Point of Delivery (POD) configuration.

As shown in Figure 7, in case of direct coupling, the maximum power absorbed is around 500 MW, while the maximum power injected is around 360 MW. The maximum power is reached in "demand mode", and its value is higher than 100 MW. This means (considering Table 2) that the connection shall be performed at the highest available voltage level with the implementation of an in-and-out single bus-bar scheme with bypass. In case of indirect coupling instead, we can see that there is a peak of injected power of around 650 MW, which is higher than the peak of absorbed power, so the connection has to be performed at the highest available voltage level with the implementation of an in-and-out double bus-bar scheme (Figure 8). Of course, also the radial configuration can be adopted if allowed by the distance of the station.

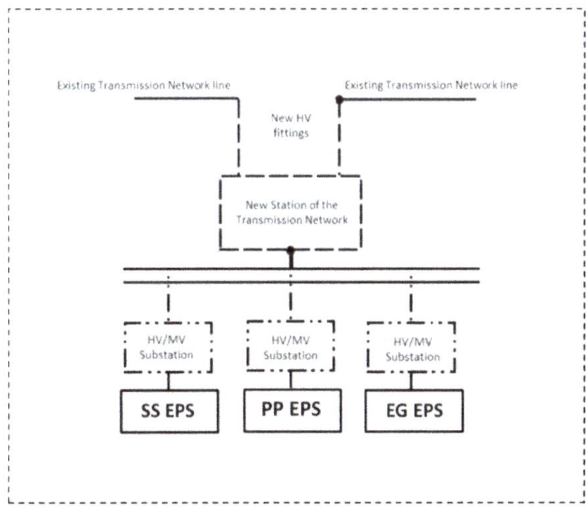

Figure 8. In-and-out double bus-bar connection scheme in case of single POD.

With the implementation of a single POD, the node cannot be seen from the external grid point of view as a proper generation node. This means that the current legislation on the connection of generators (RfG code) to the grid cannot be applied. Moreover, this configuration is as easy to implement as it is dangerous, both for the operation of the facility and of the network. In fact, the high-power spikes absorbed by the converters to feed the PP EPS loads would jeopardize the functioning of the generator, since it would be the nearest source of power. This implies high transient electromechanical torque applied to the shaft of the TG, whose integrity could be seriously compromised, both for the magnitude and for the cyclicality of the resistive torque. To avoid this situation, an electrical ESS could be implemented upstream the PP EPS, to limit the power derivatives.

4.2. Double POD with EG-Dedicated Node

It is clear that the generator should be decoupled as much as possible from the PP EPS supply. One way to do so is by implementing a double POD solution, with one POD dedicated to the EG EPS and the second one for the SS EPS and PP EPS. Figure 9 shows the active power profile at the POD for the SS EPS plus PP EPS, and Figure 10 shows the generation profile of the EG EPS.

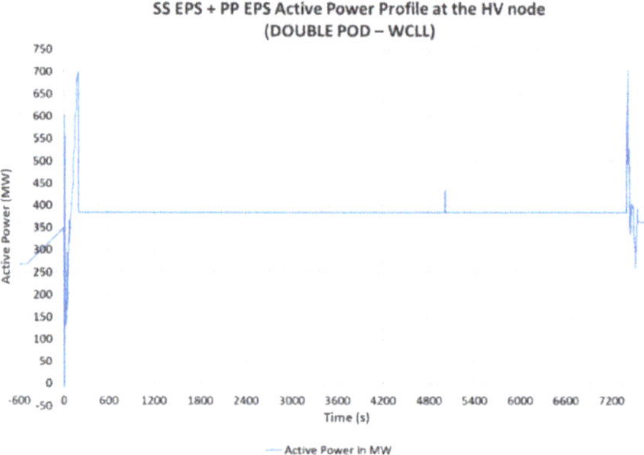

Figure 9. Power profile at the Steady State Electrical Power System (SS EPS) + Electrical Power System (PP EPS) HV node in case of double POD configuration.

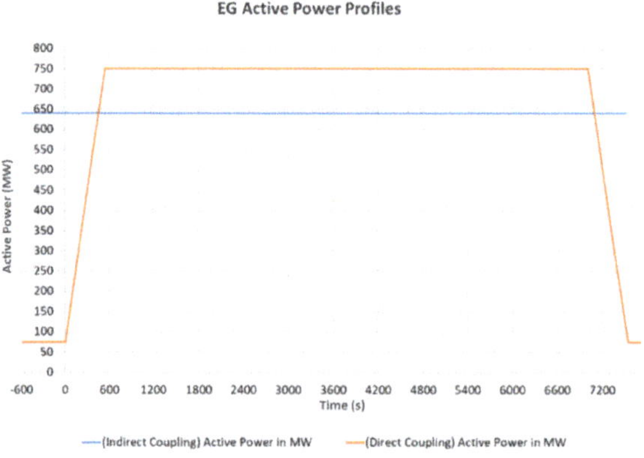

Figure 10. Power output of the generator in case of direct and indirect coupling.

For convention, the power supplied by the external grid is represented as positive, while the power injected into the grid is negative. However, in Figure 10 the generator output power profile is defined from the generator point of view, so it is represented as positive. Of course, from the external grid point of view, that power is negative, since it is injected into the grid itself.

Regarding the SS EPS plus PP EPS POD, since the maximum power required from the grid is around 700 MW, the connection shall be performed at the highest voltage level available, with the implementation of an in-and-out single bus-bar scheme with bypass (see Table 2). Concerning the EG dedicated POD instead, the maximum power injected is 750 MW in case of direct coupling and 640 MW in case of indirect coupling. Therefore, independently from the coupling configuration, the connection shall be performed at the highest voltage level available, with the implementation of an in-and-out double bus-bar scheme.

Figure 11 reports the in-and-out connection schemes in case of double POD configuration with EG-dedicated node.

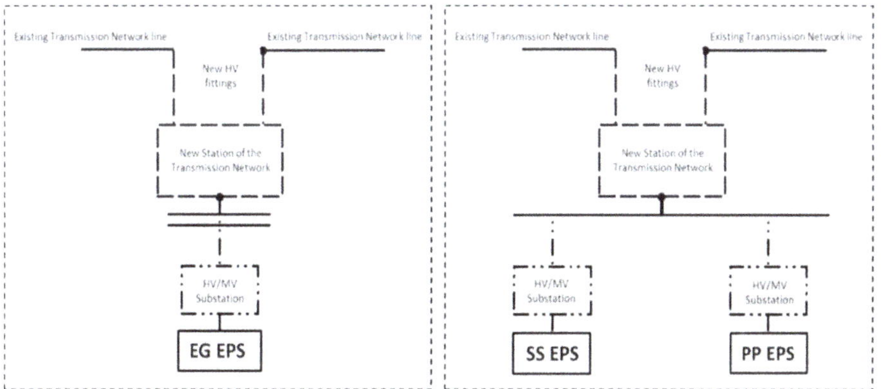

Figure 11. In-and-out connection scheme in case of double POD with EG-dedicated node.

As long as the generation node is well defined, all the requirements for the connection and operation of PGMs in the grid can be applied. Assuming the two POD to be fed by stations of the Transmission Network that are far enough from each other (in an electrical sense), in this configuration, the generator could operate more safely. Moreover, it could operate in line with the prescriptions for its specific category, bearing in mind the intrinsic limitations of the plant evaluated in Section 3.

Of course, from the demand point of view, the influence of the pulsed loads in this case is completely reflected to the transmission system.

4.3. Double POD with PP EPS-Dedicated Node

Another possible solution is the implementation of a double POD with one node dedicated to the PP EPS. Regarding the SS EPS plus EG EPS node, Figure 12 shows that, both in case of direct and indirect coupling, the maximum power is attained in generation mode, so the requirements to apply for the connection are those referred to the generation nodes.

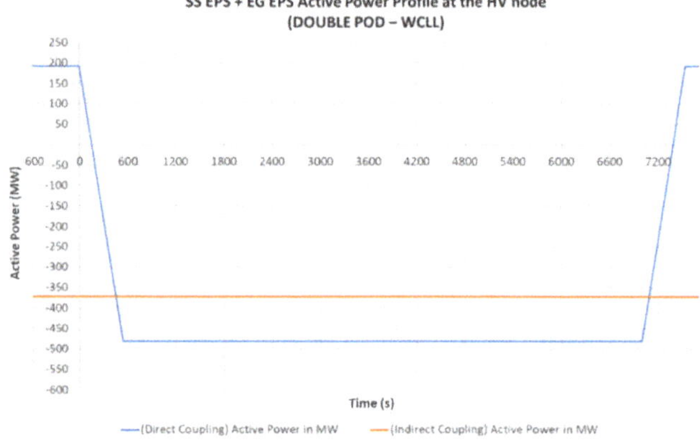

Figure 12. Power profiles at the SS EPS + EG EPS HV node in case of double POD configuration.

In particular, since the maximum power injected is higher than 350 MW, the connection shall be implemented at the highest available voltage level, with the in-and-out double bus-bar scheme (Table 2). In case of indirect coupling, the node appears to grid as a simple generation node, since the power needed for the SS EPS is entirely supplied by the EG EPS during all the operational phases. Therefore, of course the net output at the HV node level is lower than the previous configuration (EG-dedicated node), but the requirements for the connection and operation of PGMs in the grid are still applicable. In case of direct coupling instead, the node is both a generation and demand node since the EG EPS can only supply the whole power required from the SS EPS during the burn time.

Concerning the PP EPS-dedicated POD, the preliminary power profile at the HV node is presented in Figure 13.

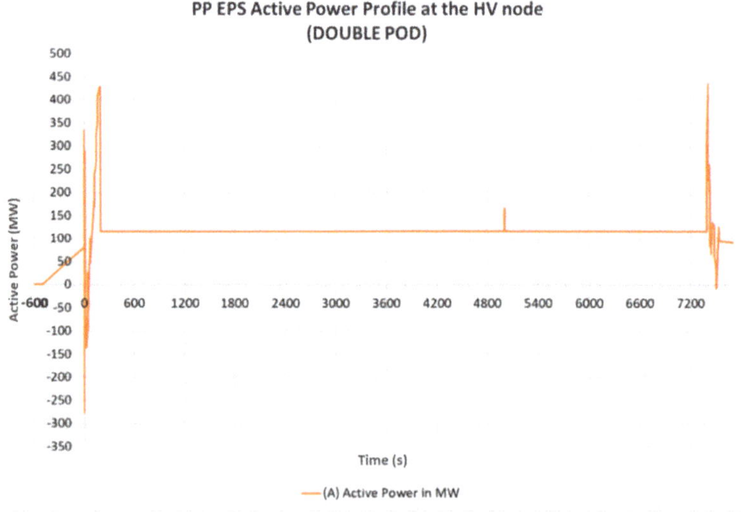

Figure 13. Power profile at PP EPS-dedicated HV node in case of double POD configuration.

The maximum power absorbed at the node level is higher than 100 MW, so the connection shall be implemented at the highest voltage level available with the in-and-out single bus-bar scheme with bypass.

Figure 14 reports the in-and-out connection schemes in case of double POD configuration with PP EPS-dedicated node.

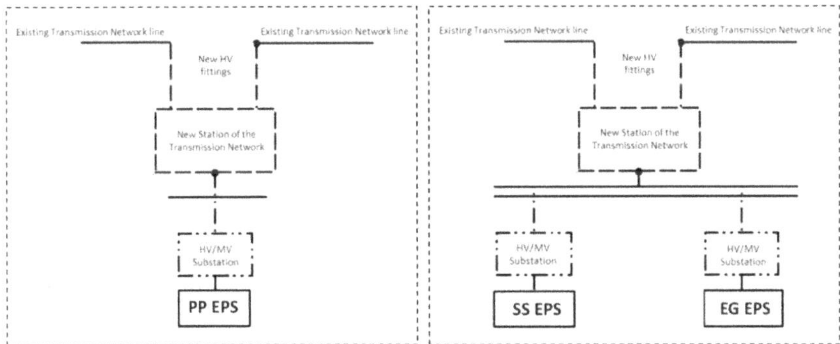

Figure 14. In-and-out connection scheme in case of double POD with PP EPS-dedicated node.

This configuration allows preserving the advantages of the previous case, decoupling both the SS EPS and the EG EPS from the supply of the PP EPS. Since the SS EPS requires practically constant input power, the generator can operate safely, supplying both the internal SS power system and the grid without further stresses and, in case of indirect coupling, as a conventional PGM connected to the grid. At the same time, the power peaks detected by the grid at the PP EPS POD, are of course lower than those related to the previous case where the SS EPS contribution makes them reach higher values.

5. Features of the Point of Delivery (POD)

This section reports further evaluations on the Point of Delivery (POD) that can be performed starting from the preliminary power profiles elaborated through DIgSILENT PowerFactory [13]. Indeed, talking about the active power profiles, some data can be extrapolated in order to have an idea of the main electrical features of the HV node (or nodes) to which DEMO should be connected. Concerning the reactive power instead, some general considerations are reported.

5.1. Power-Frequency Control

The requirements in terms of frequency control and admissible frequency variations are defined in the ENTO-E document "P1—Policy 1: Load-Frequency Control and Performance" [14]. The power-frequency characteristic of a power system is linked to its capability of limiting the frequency variation during an event of unbalance between generation and demand. The power-frequency variation is expressed in MW/Hz, so it physically represents the value of power variation in MW that causes a frequency variation of one Hz. Policy 1 [14] defines minimum and average power-frequency control characteristics in Continental Europe, which are respectively 15 GW/Hz and 19.5 GW/Hz. Other important values of the Policy 1 [14] that we have to consider for the purpose of this analysis are the "minimum instantaneous frequency after a loss of generation" and the "maximum instantaneous frequency after a loss of load", which are respectively 49.2 Hz and 50.8 Hz. This means that the maximum frequency deviation accepted in Continental Europe's synchronous area is ±800 mHz from the nominal value. However, it has to he stressed out that these values refer to the loss of a load or of a generation node; they do not refer to the ordinary operation of the grid. Indeed, to have a more reliable limit we should consider the maximum permissible quasi-steady-state frequency deviation that is set to ±200 mHz. This frequency deviation also represents the limit for which all the available primary control reserves are expected to be fully activated. In fact, this value represents the maximum value that can be managed only relying on the primary frequency control.

Regarding our specific case, analyzing the active power profiles, it is possible to identify several power steps, mainly due to the functioning of the magnets system and of the AH devices. In particular, the most severe event in this sense is a 426 MW active power step that occurs during the Plasma Ramp-up phase. The event, reported in Figure 15, is caused by the PP EPS operation.

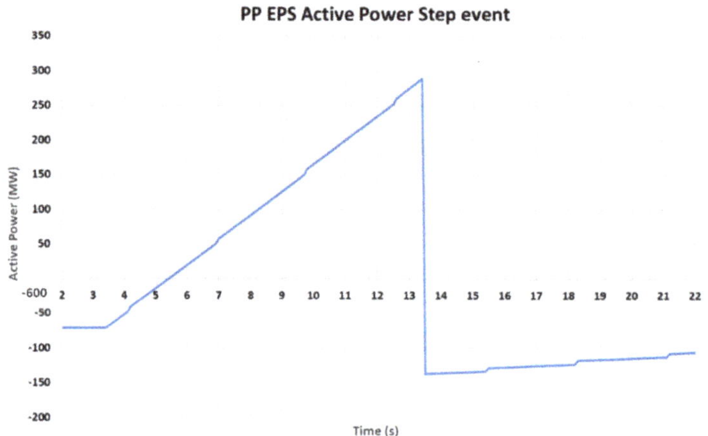

Figure 15. Active Power step required by the PP EPS during the Plasma Ramp-up phase.

Since this power step is caused by the PP EPS, it will be present in each configuration we considered in the previous section for the connection of DEMO to the electrical grid. Therefore, independently from the solution adopted, we have to evaluate the effect of this event on the grid. Considering the average value of power-frequency control characteristic for Continental Europe, we can calculate the frequency deviation caused by the step in Figure 15, which turns out to be around +21.85 mHz, significantly lower than the 200 mHz limit. Indeed, the 200 mHz limit is reached in case of what is called Reference Incident inside the ENTSO-E documents, i.e., an unbalance between generation and demand of 3 GW. Nonetheless, this event refers to an unbalance condition distributed inside the European transmission grid and not to an unbalance in a single node. In this sense, the effect of a 426 MW active power on the grid step must not be underestimated just because it theoretically does not generate a significant frequency deviation from the nominal value. Moreover, it must be stressed out that this step is not occasional but is cyclically repeated during DEMO operation, so this is another critical aspect.

5.2. Voltage Drop and Short-Circuit Power

Considering the reactive power, the connection node should be characterized through the short-circuit power. The link between the reactive power and the short-circuit power of the node is the voltage drop. It is well known that, for a fixed value of reactive power required by the facility, a higher short-circuit power of the connection node involves a lower voltage drop at node level. A first approximation of the voltage drop, expressed in per unit, ΔV [p.u.] caused by a variation of reactive power ΔQ on a node with a short-circuit power S_{sc} is provided by the well-known formula ΔV [p.u.] = $\Delta Q/S_{sc}$.

Since for now there is not any reliable data regarding the actual reactive power required for the operation of DEMO (mainly for what concerns the supply of the pulsed loads), we cannot really define a specific value for the short-circuit power needed at the connection node. Nevertheless, we can adopt an opposite approach starting from a reasonable value for the short-circuit power and from the requirements in terms of voltage drop defined by ENTSO-E. Bypassing the requirements in terms of power factor and so assuming DEMO to be a non-conventional demand facility, we can preliminarily estimate a reasonable maximum value of reactive power that the facility can demand from the grid.

Considering the studies carried out on ITER [15] and the size of DEMO with respect to its predecessor, the new facility is likely to be connected to a node with a short-circuit power rating in the order of at least 30–40 GVA. This value seems reasonable considering the configuration of the European electrical grid and also considering that nodes with this rating can be also found in Italy [16], where the shape of the country does not facilitate the creation of highly meshed grids (which means higher reliability and also higher short-circuit power).

Bearing in mind this range of values for the short-circuit power, we must define a range of values for the voltage drop that do make sense for the specific case. Considering that most of the reactive power will be required by the PP EPS (due to the power conversion systems), an estimation based on a maximum voltage drop at the connection node, around $\pm\,3 \div 5\%$, should be conservative and reliable. Indeed, the 3% value for the voltage drop is the one assumed for the Pulsed Power Electrical Network (PPEN) in ITER, which corresponds to the PP EPS in our case, as mentioned in the System Requirements Documents of ITER (available in the ITER private database). Due to the considerably greater size of DEMO, it makes sense to assume a wider range for the voltage drop.

Defining the worst case in terms of short-circuit power and voltage drop, i.e., 30 MVA and 3%, respectively, the resulting maximum reactive power demand from DEMO facility is around 900 MVAr.

Unlike the considerations reported for the active power related aspects, in this case the POD configuration affects the evaluations. As already mentioned, the main concerns in terms of reactive power absorption come from the power electronic devices needed to supply the magnets system and the AH and so the pulsed loads. Therefore, if the PP EPS is isolated from the other substations, which means we are adopting the third configuration (see Section 4.3), we can consider only its node as an unconventional demand node, and we can cut the contribution of the SS EPS to the total reactive power required from that specific node. If the first configuration is considered instead (see Section 4.1), since the EG EPS could not be operated as a conventional PGM in any case, the possibility of using it for the internal reactive power compensation is not to be excluded, still bearing in mind the drawbacks of this solution.

6. Conclusions

Considering the state of the art of the DEMO project, the present work has been carried out with the purpose of providing some cause for reflection on the choices that have been made and that will be made; it is not meant to define final design solutions. In fact, it is important to stress out that at this stage of the design, the data is continuously questioned, so the analyses carried out for the realization of this article aim at providing guidelines, procedures and addresses.

The data available from the studies that are being performed in research centers all over Europe allowed a first estimation of the input and output power profiles of the facility. Through the analysis of these profiles, it has been possible to evaluate the limitations of the facility, both in case of direct and indirect coupling between the PHTS and the PCS, from the generator operation point of view. In this sense, the conclusion is that the direct coupling configuration, despite being the simplest from the constructive point of view, would lead to the impossibility of considering DEMO as a conventional power plant, unless significant adjustments are foreseen. In any case, the eventuality that DEMO could be considered as a non-conventional generation and demand facility is not to be excluded. In this sense, no matter which configuration will be selected, the European grid could allow the operation of the facility. However, this would not be in line with the purpose of the whole roadmap, since it would lead to a single facility that is allowed to operate freely in the grid. If designed on these terms, DEMO would not be a model for real fusion power plant prototypes; it would be a model for a new generation of non-dispatchable generation plants.

Regarding the connection configuration and characteristics, this article presented three possible solutions. In particular, the solution that allows for the implementation of an EG-dedicated point of delivery, even if it is theoretically not in line with the idea at the base of DEMO, which is the self-sustainment of its loads, is the one that shows more advantages. From the grid connection point of view, the generator could be operated in line with the requirements for its category, being the node a pure generation node. From the facility point of view instead, the turbine-generator group would be decoupled from the power spikes absorbed and injected by the PP EPS, guaranteeing a longer life of the mechanical components.

Author Contributions: Conceptualization, M.C.F. and M.P.C.; methodology, M.C.F. and M.P.C.; software, M.P.C.; validation, M.C.F., A.L. and S.C.; formal analysis, M.C.F. and M.P.C.; investigation M.P.C.; resources, A.L. and S.C. data curation, M.P.C. and A.L.; writing—original draft preparation, M.P.C.; writing—review and editing, M.P.C. and M.C.F.; visualization, M.P.C.; supervision, M.C.F., A.L. and S.C.; project administration, M.C.F., A.L. and S.C.; funding acquisition, M.C.F., A.L. and S.C. All authors have read and agreed to the published version of the manuscript.

Funding: This work has been carried out within the framework of the EUROfusion Consortium and has received funding from the Euratom Research and Training Programme 2014–2018 and 2019–2020 under grant agreement No 633053. The views and opinions expressed herein do not necessarily reflect those of the European Commission.

Conflicts of Interest: The authors declare no conflict of interest.

References

1. EUROfusion, European Research Roadmap to the Realization of Fusion Energy. 2018. Available online: https://www.euro-fusion.org/eurofusion/roadmap/ (accessed on 30 March 2020).
2. ITER. What Will ITER Do? Available online: https://www.iter.org/sci/Goals (accessed on 30 March 2020).
3. Barucca, L.; Ciattaglia, S.; Chantant, M.; Del Nevo, A.; Hering, W. Status of EU DEMO Heat Transport and Power Conversion Systems. *Fusion Eng. Des.* **2017**, *136*, 1557–1566. [CrossRef]
4. Morris, J.; Kovari, M. Time-dependent power requirements for pulsed fusion reactors in system codes. *Fusion Eng. Des.* **2017**, *124*, 1203–1206. [CrossRef]
5. Federici, G.; Bachmann, C.; Barucca, L.; Biel, W.; Boccaccini, L.; Brown, R.; Bustreo, C.; Ciattaglia, S.; Cismondi, F.; Coleman, M.; et al. DEMO design activity in Europe: Progress and updates. *Fusion Eng. Des.* **2018**, *136*, 729–741.
6. Ciattaglia, S.; Federici, G.; Barucca, L.; Lampasi, A.; Minucci, S.; Moscato, I. The European DEMO Fusion Reactor: Design Status and Challenges from Balance of Plant Point of View. In Proceedings of the 17 IEEE International Conference on Environment and Electrical Engineering (EEEIC 2017), Milan, Italy, 6–9 June 2017.
7. Lampasi, A.; Minucci, S. Survey of Electric Power Supplies Used in Nuclear Fusion Experiments. In Proceedings of the 17 IEEE International Conference on Environment and Electrical Engineering (EEEIC 2017), Milan, Italy, 6–9 June 2017.
8. ENTSO-E. *COMMISSION REGULATION (EU) 2016/631 of 14 April 2016 Establishing a Network Code on Requirements for Grid Connection of Generators*; ENTSO-E: Brussels, Belgium, 2016.
9. ENTSO-E. *Limited Frequency Sensitive Mode, ENTSO-E Guidance Document for National Implementation for Network Codes on Grid Connection, 31 January 2018*; ENTSO-E: Brussels, Belgium, 2018.
10. ENTSO-E. *Frequency Sensitive Mode, ENTSO-E Guidance Document for National Implementation for Network Codes on Grid Connection, 7 November 2017*; ENTSO-E: Brussels, Belgium, 2017.
11. Fitzgerald, A.; Kingsley, C.; Umans, S. *Electric Machinery*; McGraw-Hill Education; McGraw-Hill: York, NY, USA, 2002; Chapter 5.5; pp. 293–294.
12. Terna, Guida Agli Schemi di Connessione. *Attachment A2 of Terna Network Code, July 2015*; Terna: Rome, Italy, 2015. Available online: https://download.terna.it/terna/0000/0105/19.pdf (accessed on 30 March 2020).
13. DIgSILENT. Load Flow Analysis. Available online: https://www.digsilent.de/en/load-flow-analysis.html (accessed on 30 March 2020).
14. UCTE. *Continental Europe Operation Handbook P1, Load-Frequency Control and Performance*; UCTE: Brussels, Belgium, 2009.

15. Benfatto, I. Power converters for ITER. *Comput. Sci* **2006**, 231–247. Available online: http://cds.cern.ch/record/987554/files/p231.pdf?version=1 (accessed on 30 March 2020).
16. Terna. *Qualità del Servizio di Trasmissione, Valori Minimi e Massimi Convenzionali Della Corrente di Cortocircuito e Della Potenza di Cortocircuito Della Rete Rilevante con Tensione 380-220-150-132 kV*; Terna: Rome, Italy, 2017.

Article

Electrical Loads and Power Systems for the DEMO Nuclear Fusion Project

Simone Minucci [1], Stefano Panella [2], Sergio Ciattaglia [3], Maria Carmen Falvo [2] and Alessandro Lampasi [4,*

[1] Department of Economics, Engineering, Society and Business Organization, University of Tuscia, 01100 Viterbo, Italy; simone.minucci@unitus.it
[2] Department of Astronautics, Energy and Electrical Engineering, University of Rome Sapienza, 00184 Rome, Italy; stefano.panella@uniroma1.it (S.P.); mariacarmen.falvo@uniroma1.it (M.C.F.)
[3] EUROfusion Consortium, 85748 Garching bei München, Germany; sergio.ciattaglia@euro-fusion.org
[4] National Agency for New Technologies, Energy and Sustainable Economic Development (ENEA), 00044 Frascati, Italy
* Correspondence: alessandro.lampasi@enea.it

Received: 16 March 2020; Accepted: 22 April 2020; Published: 4 May 2020

Abstract: EU-DEMO is a European project, having the ambitious goal to be the first demonstrative power plant based on nuclear fusion. The electrical power that is expected to be produced is in the order of 700–800 MW, to be delivered via a connection to the European High Voltage electrical grid. The initiation and control of fusion processes, besides the problems related to the nuclear physics, need very complex electrical systems. Moreover, also the conversion of the output power is not trivial, especially because of the inherent discontinuity in the EU-DEMO operations. The present article concerns preliminary studies for the feasibility and realization of the nuclear fusion power plant EU-DEMO, with a special focus on the power electrical systems. In particular, the first stage of the study deals with the survey and analysis of the electrical loads, starting from the steady-state loads. Their impact is so relevant that could jeopardy the efficiency and the convenience of the plant itself. Afterwards, the loads are inserted into a preliminary internal distribution grid, sizing the main electrical components to carry out the power flow analysis, which is based on simulation models implemented in the DIgSILENT PowerFactory software.

Keywords: balance of plant; DEMO; electric loads; nuclear fusion; plasma; power flow; power supply; power systems

---

1. Introduction

EU-DEMO (the DEMOnstration fusion power reactor proposed by the European Union), or simply DEMO, is a unique European project, as it will be the first demonstrative nuclear fusion power plant able to produce and distribute electrical power throughout Europe, thanks to a connection with the European High Voltage (HV) electrical grid (typically at 400 kV) [1–3].

To accomplish this challenging purpose, the European Union set up the EUROfusion Consortium, whose main goals and tasks are summarized in the "European Research Roadmap to the Realisation of Fusion Energy" [4]. The schedule and the milestones of the Roadmap are sketched in Figure 1.

Even though other alternative approaches are being investigated in EUROfusion [4] and in another research facility [5], the EUROfusion Roadmap is based on two tokamak projects: DEMO and ITER [6]. The latter is currently under construction in Cadarache (France) with a worldwide contribution and aims at:

- Producing 500 MW of fusion power for pulses of at least 400 s.

- Demonstrating the integrated operation of technologies for a fusion power plant.
- Achieving a deuterium-tritium plasma where reactions are sustained through internal heating.
- Testing tritium breeding.
- Demonstrating the safety characteristics of a fusion device, both for human and environment.

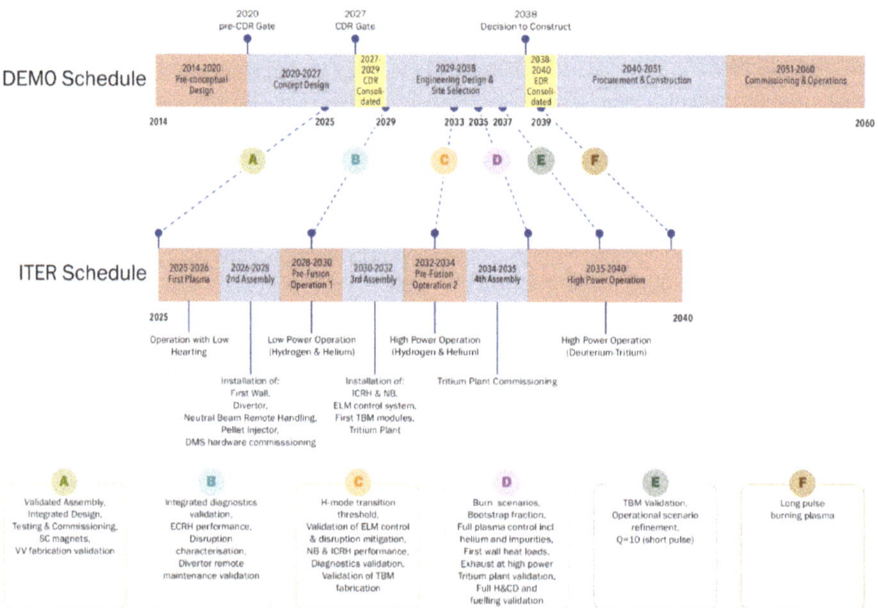

Figure 1. Overview of the European Roadmap to the realization of Fusion Energy [4]. The reported dates are still indicative.

These goals are important and partially common with the DEMO project, so that it can use ITER-like solutions for possible future issues. However, DEMO will be bigger than ITER in terms of size and of required services, also because it will be connected to the grid to deliver the produced electrical energy, unlike ITER. Also the time required for the realization of these two projects is different: while it is foreseen to start ITER first experiments by 2025 and to operate with deuterium and tritium by 2035, DEMO is expected to be in operation around by 2050.

In the past, fusion devices were generally not regarded as nuclear facilities and did not need a nuclear license. However, ITER and DEMO are much more critical in terms of tritium inventory, neutron flux, pulse duration, stored magnetic energy, cooling system enthalpy and amount of helium at 4 K. ITER demonstrated its safety and obtained the nuclear license to start the construction. Nevertheless, specific nuclear regulations are likely to be introduced for next-generation devices, also depending on the host country. As DEMO is expected to have more neutronic flux and more inventory of tritium than ITER, a license from authorities will be necessary before starting the construction of safety-classified systems.

As a nuclear facility, a specific design is necessary for the DEMO Balance of Plant (BoP), that is the nuclear engineering term referred to all the supporting and auxiliary systems needed for energy conversion and delivering, excluding all the nuclear components. Therefore, EUROfusion is promoting a multidisciplinary research and engineering activity that is approaching the design of the DEMO BoP [7,8], even moving from the experience of the other experimental tokamaks and nuclear-fission power plants. One of the most critical part of the BoP is the electrical power system, also because this system is not trivial both in terms of size and of complexity.

The present paper introduces the preliminary studies for the feasibility and realization of the electrical system of the nuclear fusion power plant DEMO. In particular, the results of the following activities are described:

- Characterization and classification of the expected of the electrical loads belonging to the specific DEMO systems by analyzing literature data, technical specifications and design progresses of DEMO itself and of previous tokamak plants, especially ITER and its satellites JT-60SA [9] and DTT [10].
- Preliminary design of part of the internal distribution network and preliminary sizing of the main electrical components of part of the internal distribution grid (conception and development of simulation models for the power flow analysis) then implemented in the DIgSILENT PowerFactory software environment.

The loads characterization and classification is the starting point for first electrical designs and is expected to provide more realistic data than those in previous analyses based on theoretical considerations [11].

This paper is organized in seven sections. Section 2 introduces the relevant DEMO figures and explains possible configurations and operation phases. Section 3 presents the preliminary layout of the DEMO site as used for the electrical analysis. Section 4 introduces the main options of the BoP and the basic principles for the design of the electrical systems. Section 5 is focused on the survey of the main DEMO subsystems and electrical loads and on the results obtained by the load analysis and characterization. Section 6 presents the results on the preliminary design and sizing of a part of the internal distribution grid. Section 7 reports the paper's conclusions.

2. DEMO Characteristics and Operation Phases

The tokamak operations are based on the heating of a plasma up to temperatures at which it is self-sustained by the fusion processes induced by the ion thermal motion. In DEMO, the plasma heats up the surrounding structure, the tokamak Breeding Blanket (BB), and the fluid used to cool down the BB can drive a Turbine Generator (TG) through proper heat exchangers. The energy produced by such process is expected to be higher than the energy employed to initiate the fusion reactions. In order to reach high temperatures (about 150 M°C), DEMO could use three different kinds of additional Heating and Current Drive (H&CD) systems [10,12]:

- Radiofrequency heating at ion cyclotron resonance frequency (ICRH), based on high-intensity beams of electromagnetic radiation able to excite the ions in the plasma at their own resonance frequency (in the order of tens of megahertz).
- Microwave heating at electron cyclotron resonance frequency (ECRH), based on electromagnetic radiations able to heat the electrons (and then the ions) up in the plasma at their own resonance frequency (in the order of hundreds of gigahertz).
- Negative neutral beam injectors (N-NBIs or simply NBIs), shooting neutral high-energy particles into the plasma where they transfer their energy by collisions.

The heat produced by the plasma fusion must be transferred to Power Conversion System (PCS) able to transform it into electrical energy with maximum possible efficiency and reliability: this is the main scope of the BoP. In the present status of the DEMO project, two alternative solutions are considered as basic fluid to cool down the BB: water and helium [8,13,14]. Consequently, two options are considered for the Primary Heat Transfer System (PHTS):

- Water Cooled Lithium Lead (WCLL).
- Helium Cooled Pebble Bed (HCPB).

The DEMO operations are strictly related to the physics of the plasma. This also implies a difficulty to achieve very long or steady-state operations that would be preferable for the energy budget. A lot of fusion research is devoted to the possibility of steady-state operations, but presently without relevant practical results. On the other hand, pulsed operations simplify tokamak physics and may be more flexible in an energy market ruled by renewable sources. Presently, the DEMO operations are supposed to be pulsed in basic option but could be steady-state in future advanced ones. Therefore, unlike nuclear-fission power plants, the DEMO BoP and electrical systems must be designed for pulsed operations for both the two cooling system options in the PHTS-PCS.

In fact, even though a relevant output power is produced only during the plasma flat-top phase, other operation phases are necessary to achieve the correct execution of the pulse. In particular, the DEMO operations consist of seven phases:

(1) Central solenoid (CS) pre-magnetization. Some superconductive magnets, in particular those located in the toroid hole (CS), must be energized to attain a suitable value of magnetic flux into the Vacuum Vessel (VV). The energization current in the DEMO CS modules is expected to be up to 45 kA and to span the range ±45 kA [15]. The duration of the pre-magnetization phase mainly depends on the maximum possible charge rate of the superconductors. The pre-magnetization is typically executed by supplying the superconductors at rather constant voltage, producing an increasing current and resulting in an increasing power demand from the grid. The start of the pre-magnetization phase is conventionally set to a negative value of time and ends at zero.

(2) Plasma breakdown. It is the shortest (about 1 s) but also the most critical phase from the power supply point of view. The plasma initiation requires high-power pulses in several superconducting coils (at least in the CS). The effective presence of the plasma in the tokamak vessel starts in this phase.

(3) Plasma ramp-up. In this phase, the plasma current is progressively increased by the coils and H&CD sources. The ramp-up must be slow in order to keep the plasma under control. On the other hand, the available CS magnetic flux is consumed by the ramp-up duration, reducing the useful time for the production of electrical energy.

(4) Heating flat-top. All the H&CD sources are used to heat up the plasma until fusion temperature and conditions are reached.

(5) Burn flat-top. DEMO produces energy thanks to the nuclear fusion reactions. Ideally, during this phase, the fusion reactions taking place at a sustainable rate guarantee the self-sustainment of the plasma. This phase is very long, about 7200 s (2 h).

(6) Plasma ramp-down. The plasma is gradually switched off, maintaining its control by coil and H&CD power.

(7) Dwell time. It is the time required after the pulse to bring DEMO to a condition stable enough to start a new pulse and to create an adequate vacuum inside the plasma chamber. Of course, it is desirable that the DEMO design will progress towards negligible dwell times. Actually, dwell time conventionally includes also the CS pre-magnetization phase since the CS charging action belongs to those operations necessary to complete a pulse and start the following one. However, since this operation does not last the whole dwell time, in this work they are managed as two different and consecutive phases.

This pulsed behavior may introduce specific problems for the BoP. First, discontinuous operations would be damaging to the turbine, then the variable flow of the expected huge powers may let some instabilities arise into the external grid that could even refuse or limit the connection. This problem is even more critical because of the relevant reactive components in power. The durations of the main plasma phases according to last DEMO design are summarized in Table 1.

Table 1. Summary of the DEMO operation phases with typical durations.

Phase	Phase Name	Typical Time Duration
1	CS pre-magnetization	500 s
2	Plasma Breakdown	1.4 s
3	Plasma ramp-up	180 s
4	Heating flat-top	10 s
5	Burn flat-top	7200 s
6	Plasma ramp-down	150 s
7	Dwell time	200 s
Total time duration of the DEMO operations		≈8240 s

In order to reduce the output power fluctuations, an intermediate buffer system could be inserted between the PHTS and the PCS. Therefore, two different approaches are under evaluation about the PHTS-PCS coupling [13,14] for both the two cooling options:

- PHTS-PCS direct cycle, characterized by direct coupling between the PHTS and the PCS.
- PHTS-PCS indirect cycle, having an Intermediate Heat Transfer System (IHTS) and thermal Energy Storage System (ESS) between the PHTS and the PCS.

The thermal efficiency in the direct cycle option is higher but the TG's life cycle is compromised because it is turned on only during the burn flat-top phase. The thermal energy is extracted from the BB by water or helium, then it is transferred to steam supplying the steam turbine.

In the indirect cycle, the PHTS is coupled with the PCS through the IHTS based on a molten-salt ESS. Its task is to store thermal power during the burn flat-top phase (removed by the molten salt) and its delivery to the PCS. In this way, the TG can operate almost in steady-state at 80% of the PHTS-rated power without interruption nor fluctuations during the plasma phases when fusion heat is not available for energy conversion. Therefore, it can produce almost constant electrical power and rotate at a rather constant speed, thus avoiding thermo-mechanical cycling issues.

Considering the two coupling configurations and the two coolant options, the DEMO BoP could be based on one out of the following four different possible configurations, as summarized in Figure 2:

(1) WCLL in direct cycle (Figure 2a).
(2) WCLL in indirect cycle (Figure 2b).
(3) HCPB in direct cycle (Figure 2c).
(4) HCPB in indirect cycle (Figure 2d).

It is worth noticing the presence of two steam generators in Figure 2b. Unlike the HCPB configuration in Figure 2d, the IHTS in the WCLL BoP takes and stores not all the power coming from the BB PHTS but only the fraction coming from the first wall that is delivered to the PCS during the dwell time using a suitable steam generator [13].

The selection of the optimal configuration is expected to be completed in the next years basing also on the outcomes of the research on the DEMO electrical loads and the design of its distribution network.

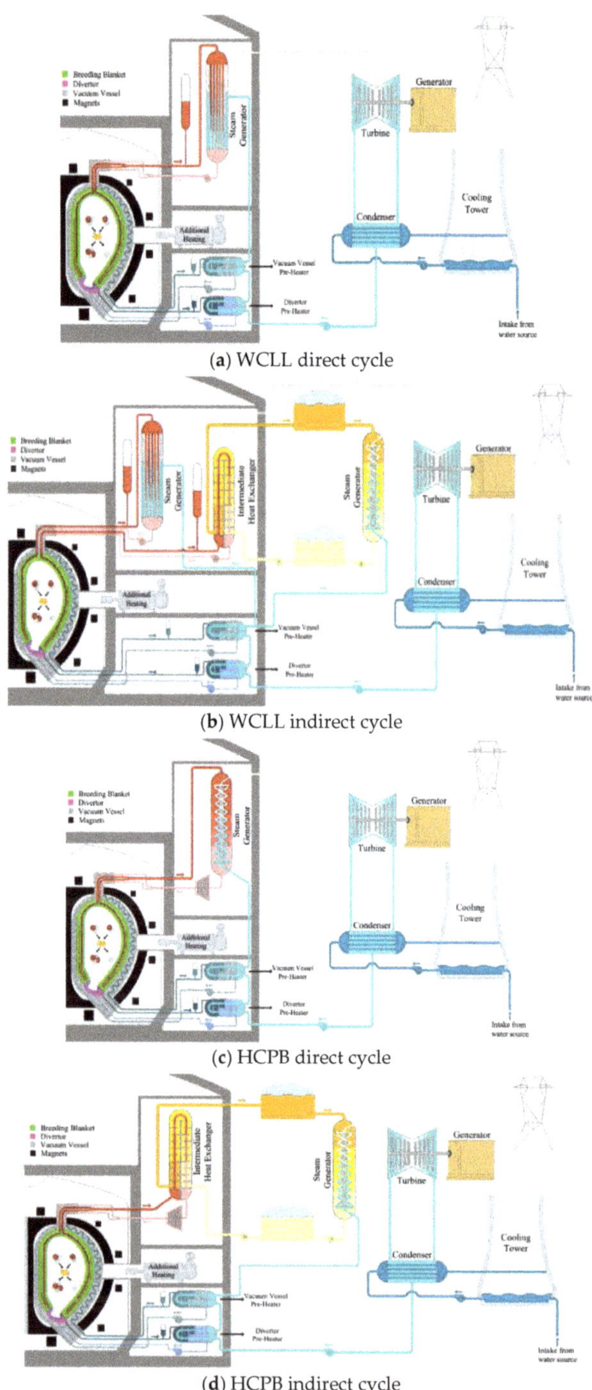

(**a**) WCLL direct cycle

(**b**) WCLL indirect cycle

(**c**) HCPB direct cycle

(**d**) HCPB indirect cycle

Figure 2. The four possible PHTS-PCS configurations that are under evaluation for the DEMO design: (**a**) WCLL direct cycle, (**b**) WCLL indirect cycle, (**c**) HCPB direct cycle, (**d**) HCPB indirect cycle.

3. Preliminary DEMO Layout

Figure 3 summarizes the preliminary layout that is presently expected for the DEMO site [7,16], mostly based on ITER's one. The actual location of the DEMO site is not yet identified and will be defined also following the outcomes of DEMO electrical analysis and requirements. Nevertheless, it is important in order to assess the electrical distribution layout, the electrical loads and cables characteristics.

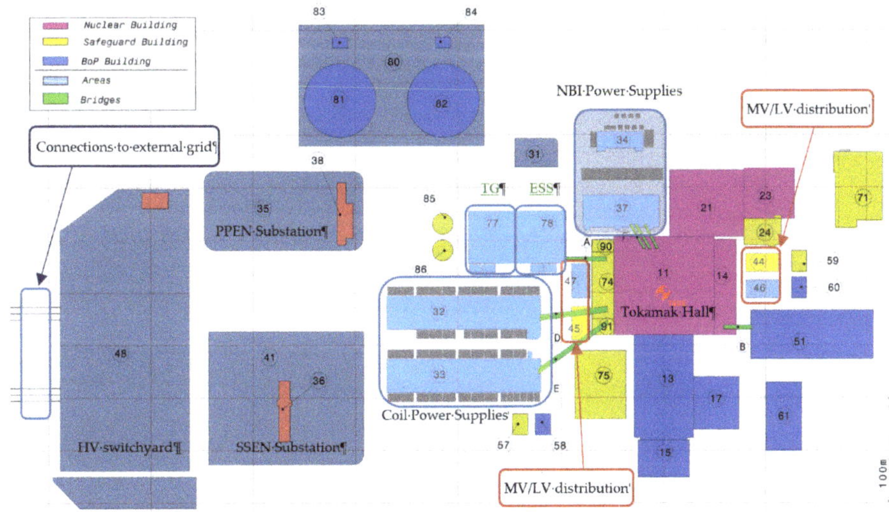

Figure 3. Preliminary layout of the EU-DEMO site.

4. Basic Principles for the Design of the DEMO Electrical Power System

DEMO power systems can be divided in three electrical distribution groups:

- Steady-State Electrical Network (SSEN), supplying all the auxiliaries, including the components relevant for nuclear safety and the protection of the investments. The steady-state loads require a rather constant power.
- Pulsed-Power Electrical Network (PPEN), supplying the pulsed loads, i.e. the superconducting coils and the H&CD systems. The pulsed loads require a time-changing power only during plasma operations.
- Turbine Generator (TG), connecting the DEMO power plant to the external grid to deliver the net electrical power produced by the PCS.

Figure 4 shows a preliminary sketch of a possible configuration of the DEMO electrical power system with its three sub-distribution groups.

The scheme in Figure 4 should be integrated by emphasizing the buses for the safety loads and by inserting the systems for the reactive power compensation and harmonic filtering. Such systems are relevant in ITER (occupying an area approximately corresponding to the DEMO Area 35 in Figure 3), but their ratings and placements can be defined in DEMO only after an adequate survey of the electrical loads.

Because of the deeply different nature of the electrical loads connected at SSEN and PPEN distributions, the former can be exhaustively characterized into an electrical load list, as described in the rest of the paper, but the latter need further specifications in terms of time evolution of the power profiles. Such difference heavily affects also the power profiles at the Point of Delivery, requiring further design and optimization stages to meet the requirements imposed by the grid regulator.

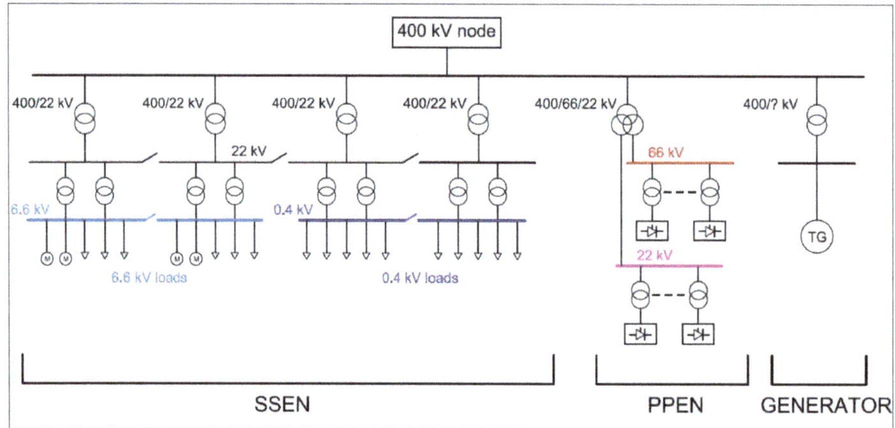

Figure 4. Preliminary sketch of a possible DEMO electrical power system configuration, emphasizing the division of the electrical distribution into three groups: SSEN, PPEN and TG.

The amount of electrical power delivered by DEMO to the external grid depends on the selected option for the PHTS-PCS and on its control strategy. Figure 5 shows the input and output of the DEMO electric power depending of the PHTS coupling. The figure is essentially qualitative as PPEN contribution is still under analysis, but some data available for a non-optimized coil system show that it could require input powers at breakdown even higher than 1 GW with relevant reactive power components [15]. More in general, the input power requested could exceed the produced power in some time intervals.

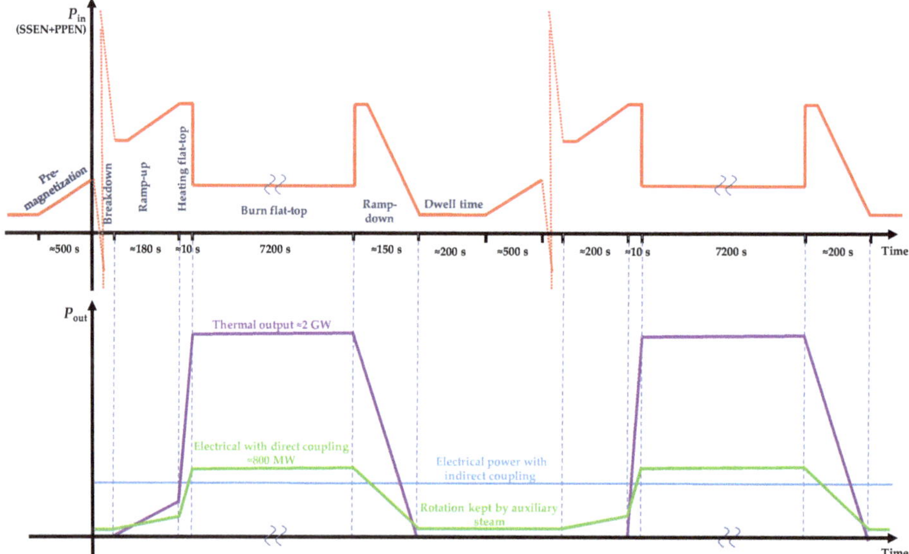

Figure 5. Qualitative (not in scale) curves of input (red line) and output powers. The thermal power produced by the plasma and fusion processes (purple line) can be transformed in an electrical power whose characteristics are different for the direct (green line) and indirect coupling (blue line).

With the direct PHTS-PCS coupling, the power for the DEMO electrical loads should be supplied only by the external grid during the dwell time between the pulses and during the ramp-up and ramp-down phases too.

In order to avoid excessive mechanical and thermal stresses on the turbine, it is supplied with a proper mass flow rate of steam which has to guarantee the turbine operation at 10% of its nominal capacity during the dwell time. In order to cover the steam production during the dwell time, the most suitable and efficient solution consists in the introduction of a small thermal ESS loop, managed in one of the following three main configurations:

- The "Steam only" configuration uses a portion of the steam in the PHTS for the molten-salt loop.
- The hybrid configuration provides for the coordinated use of the steam from the PHTS and an electric heater.
- The "Electric only" configuration foresees only the use of an electric heater.

The most efficient solution seems to be the third one due to temperature concerns [17]. Indeed, using the steam in the PHTS to heat the molten-salt hot tank does not allow reaching a temperature high enough to guarantee proper conditions for the steam turbine during the dwell time. This could result in water damage to the steam turbine.

Besides the operation during the dwell time, thermodynamic analysis show that the maximum power output of the turbine is evaluated to be around 790 MW [18]. Considering a conservative efficiency of 0.95 for the turbine-synchronous generator coupling, the nominal active power to consider for the generator is 750 MW. In literature, a common minimum power factor to operate synchronous generators results to be 0.8. This value is also adopted in some grid legislations [17] to identify the value of the reactive power that a generator should be able to supply to the grid when required, and so to build a reference capability curve. Considering this value for the power factor, the rating of the synchronous generator in terms of apparent power is around 940 MVA.

This has to be considered an approximation since the variations in the prime mover power would actually cause transient phenomena at the synchronous machine level. From the reactive power point of view instead, the nominal value that the generator could supply to the grid is around 565 MVAR, coming from the minimum power factor assumed. However, no assumptions can be made on the reactive power profile since, in general, it is dependent on the conditions of the external grid and so it is managed by the transmission system operator.

In order to find a coherent value for the output power and to assume an efficiency for the new thermodynamic cycle, it is necessary to know the thermal energy balance. Of course, this efficiency might be different from the direct coupling case because of the presence of the IHTS. Using the IHTS, the turbine operates constantly at its nominal power, increasing the overall efficiency during the whole operation (no transients, no off-design conditions). On the other hand, the presence of a more complex heat exchange system may lead to less thermal power transferred to the PCS. For these reasons, a first approximation about the power output in case of indirect coupling has been evaluated as the total energy output during one pulse, divided by the total duration of the pulse. In this way, the nominal power of the turbine has been calculated to be around 85% of the previous case. This leads to a rated active power of around 640 MW. Taking as a reference for the power factor the value adopted in the previous case, for the indirect coupling configuration, the rated apparent power of the synchronous generator ranges around 800 MVA.

5. Summary of Electrical Loads Connected to the SSEN

The DEMO subsystems are organized according to a Plant Breakdown Structure (PBS), used also for the classifications of the electrical loads. Table 2 provides for each PBS a short description, the expected power and the electrical distribution system where the PBS and its loads is connected. Since some PBSs contain both loads connected to the SSEN and to the PPEN distribution networks, they are reported twice.

Table 2. EU-DEMO Project Plant Breakdown Structure (PBS) with expected powers and connection to the distribution grid.

PBS	PBS Description	Power		Distribution
		Active (MW)	Reactive (MVAR)	
11	Magnet System	0	0	Passive
12	Vacuum Vessel (VV)	0	0	Passive
13	Divertor System	0	0	Passive
14	Blanket (HCPB) [1]	0	0	Passive
16	Blanket (WCLL) [1]	0	0	Passive
18	Limiter	0	0	Passive
20	Cryostat	0	0	Passive
21	Thermal Shields	0	0	Passive
22	Tritium, Fueling, Vacuum	12.2	7.7	SSEN
25	Tritium Extraction and Removal (HCPB) [1]	3.0	1.9	SSEN
27	Tritium Extraction and Removal (WCLL) [1]	3.0	1.9	SSEN
30	ECRH System (main power) [2,3]	125.0	60.5	PPEN
30	ECRH System (auxiliary power) [2]	6.0	2.9	SSEN
31	NBI System (main power) [2,3]	125.0	60.5	PPEN
31	NBI System (auxiliary power) [2]	6.0	2.9	SSEN
32	ICRH System (main power) [2,3]	125.0	60.5	PPEN
32	ICRH System (auxiliary power) [2]	6.0	2.9	SSEN
40	Plasma Diagnostic & Control System	6.1	3.0	SSEN
49	VV PHTS	9.7	4.7	SSEN
50	Breeding Blanket PHTS (HCPB) [1,3]	165.6	54.4	SSEN
52	Breeding Blanket PHTS (WCLL) [1]	59.4	19.5	SSEN
54	VV Pressure Suppression System (HCPB) [1]	2.3	0.0	SSEN
56	VV Pressure Suppression System (WCLL) [1]	4.6	2.9	SSEN
58	Divertor & Limiter PHTS (HCPB) [1]	19.5	12.1	SSEN
59	Divertor & Limiter PHTS (WCLL) [1]	10.0	6.2	SSEN
60	Remote Maintenance (RM) System [4]	5.0	3.1	SSEN
61	Assembly	4.6	2.2	SSEN
63	Radwaste Treatment and Storage	3.0	1.5	SSEN
70	Balance of Plant (HCPB) [1]	12.0	5.8	SSEN
72	Balance of Plant (WCLL) [1]	12.0	5.8	SSEN
80	Site Utilities	3.1	1.9	SSEN
81	Cryoplant & Cryodistribution	101.8	63.1	SSEN
82	Electrical Power Supply (main power) [3]	300.0	300.0	PPEN
82	Electrical Power Supply (auxiliary power)	21.0	10.2	SSEN
83	Buildings	54.8	26.6	SSEN
85	Plant Control System	3.6	1.7	SSEN
87	Auxiliaries	90.9	56.4	SSEN

[1] The PBSs referred to the HCPB and WCLL options are alternative: once the final configuration will be selected, only one of the two reported powers will be requested in DEMO. [2] The powers reported for the H&CD systems (PBSs 30, 31 and 32) were estimated for the reference solution with related efficiencies (see Section 5.3). [3] Since the powers absorbed by these PBSs are not constant, their peak powers are reported. While the PPEN loads (PBSs 30, 31, 32 and 82) may be very variable, the load of PBS 50 is reduced to 20% in some phases as described in Section 5.5. [4] Even if RM System is connected to the SSEN, it mainly operates during specific maintenance phases and not during plasma phases in Table 1.

Passive systems are those not absorbing any significant electrical power. However, it is worth mentioning that systems which are labelled as passive may require a small amount of power for the instrumentation and control systems. In such cases, this power is accounted in PBS 40 or 85.

Some of listed PBSs of the project are still at very preliminary stage or under design. To provide a first estimation of the electrical power absorbed by these subsystem, and to carry out a comprehensive (although preliminary) power flow analysis of the DEMO plant, a first guess based on reasonable extrapolation from the ITER loads and power profiles was used.

For each of the PBSs listed in Table 2 the information necessary to run power flow analyses were investigated, as the load type (if belonging to SSEN or PPEN distribution), the voltage level and the power absorption in terms of active power and power factor. Whenever no information coming from the design layout regarding the power factor were available, a local reactive power compensation system is supposed for each subsystem, regulating it at a value equal to 0.95.

Note that the total input power requested by DEMO cannot be estimated only by summing the entries in Table 2 because: (i) HCPB and WCLL cooling options are alternative and (ii) the power absorbed by loads belonging to PBSs 30, 31, 32, 50 and 82 are not constant for all the phases, being them pulsed. In these cases, the only peak values of the power are reported. In particular, the PPEN loads require high powers only to initiate (phases 1 and 2 in Table 1), ramp (phases 3 and 6) and heat (phase 4) the plasma, while their contribution is lower than the output power for most of the operational time (phase 5). The details for the PPEN power profiles will be developed in next years.

The following subsections provide a brief description of the electrical loads for each "non-passive" DEMO PBS supplied by the SSEN distribution.

5.1. PBS 22: Tritium, Fueling and Vacuum

The DEMO PBS 22 collects all the components belonging to the subsystems devoted to the fuel cycle (e.g., isotope separation columns, water detritiation, molecular sieve beds), to the plasma fueling (e.g., pellet injection systems) and to the vacuum generation inside the plasma chamber (e.g., motor pumps and compressors). For all these subsystems, preliminary assumptions and designs are under evaluation based on the lesson learnt from ITER layout and projects. A first guess of the total electrical power absorbed by the three subsystems is provided as 12.2 MW, from the information collected from the design progresses and from extrapolations from ITER, taking into account also the differences between the two projects and how they reflect on the operation of each subsystem.

5.2. PBS 25/27: Tritium Extraction and Removal for HCPB/WCLL

Tritium Extraction and Removal is a subsystem devoted to the unburned and unexhausted tritium extraction from the VV and its transportation to the fuel cycle system (PBS 22-1). Afterwards, it is treated, separated from its isotopes and re-injected inside the chamber to participate to the occurring fusion reactions. This system is at embryonic state and no information regarding its structure and its power absorption have been evaluated at present, both for the two cases of cooling system based on helium or water coolant. As a first guess of the power absorption, an extrapolation from ITER layout was carried out, obtaining an estimation of 3 MW.

5.3. PBS 30, 31 and 32: Heating and Current Drive (H&CD) Systems

The H&CD systems foreseen in DEMO are ECRH, ICRH and NBI. The optimal heating source mix, as well as the power to be delivered to the plasma, is still a key issue under investigation. The definition of the final parameters is scheduled by the end of the conceptual phase, foreseen by the end of 2024. Up to now, the requests in terms of plasma heating and stabilization require to deliver 150 MW of heating power to the plasma (that needs a much higher electrical power to be supplied). Three possible options are under investigation to achieve the total amount:

- The reference solution consists in evenly distributing the total power among the three sources.
- The alternative (and rather probable) solution consists in excluding ICRH and moving its power to ECRH (100 MW) without changing the NBI.
- An open option consists in producing all the necessary power by ECRH.

It is important to stress that the functions of the H&CD systems need to be supplied by two different kinds of power supplies:

- Main Power Supplies, connected to the PPEN.
- Auxiliary Power Supplies, connected to the SSEN.

In ITER the auxiliary powers of the H&CD systems are included in the respective PBSs and are accounted to be all equal to about 3 MW. As the DEMO power supplies are expected to manage higher powers, an auxiliary power of 6 MW is estimated from the SSEN for each of the three H&CD PBSs.

The issue of the time evolution and efficiencies [10,12] of the H&CD Main Power Supplies is more complex and will be addressed in a future paper by the authors. The average efficiencies for the expected technologies are foreseen to be about 40% of the power delivered to the plasma [12], resulting in a power of 125 MW for the main power of each H&CD system.

5.4. PBS 40: Plasma Diagnostics and Control Systems

Since the Plasma Diagnostic & Control System is still to be defined, the electrical power absorbed has been estimated by the ITER data. However, since the DEMO facility is expected to be developed under a more mature knowledge of fusion physics and technology with respect to ITER, its diagnostic system is supposed to be much simpler, leading to an estimation of 6.1 MW, supposing a global power factor equal to 0.9. The electrical power includes the local air conditioning of the cubicles serving the diagnostics.

5.5. PBS 50: BB Primary Heat Transfer System (HCPB)

The Primary Heat Transfer System based on the HCPB collects all the components and systems devoted to the fusion heat recovery and delivery to the PCS. The main heat extraction zone is the BB, supplying about 85% of the total fusion heat. Other fusion heat sources such as the VV, limiter and divertor are addressed in PBS 49 and PBS 58. As mentioned in Section 2, two possible layouts are under investigation to couple PHTS to the PCS, in order to mitigate the intrinsically pulsed behavior of the plasma for TG, that are the indirect (Figure 2d) and direct coupling (Figure 2c).

In the first configuration, the electrical loads are related both to the PHTS and to the IHTS. Its aim is to recover heat to be supplied to the turbine during dwell time and all phases when fusion heat is not available for energy conversion, in order to keep it constant in its operation.

In the direct cycle configuration, instead, the electrical loads are only those belonging to the PHTS system and those belonging to the small ESS.

Considering the segmentation option with 18 sectors (each of them including three outer blanket and two inner blanket segments), the HCPB BB is divided into nine independent circuits (for safety reasons, in order to limit common-mode failures), serving two sectors each. The sectors are fed by nine loops (each with two compressors), six designed for the Outboard Blanket (OB) sectors and three designed for the Inboard Blanket (IB) sectors. The main electrical load for each loop is the electrical motor connected to the compressors, for which a first power estimation is provided based on the thermo-hydraulic design of the loops themselves (9.2 MW of mechanical power per compressor with efficiency equal to 0.95).

Because of the very high electrical power absorbed by the motor compressors, resulting in a total power in the order of hundreds of megawatts, it was decided to regulate them in order to reduce the power consumption during the scenario phases when no thermal power comes from the fusion reactions. Unfortunately, the power transitions from regulated to full power and vice versa cannot be performed at any time rate because the current technology for helium compressors allows a ramp-up and ramp-down at a maximum rate of about 10–20%/minute. Also with controlled ramps, the frequent power variations induce stresses on the compressors that are justified by the relevant energy saving that can be achieved (more than 100 MW for more than 500 s). Some possible modulations of the compressor power are sketched in Figure 6. The selected modulation rate will be a trade-off between energy saving and components stress. Presently, the assumed tradeoff is 20%/min but faster rates will be investigated also considering technology progresses.

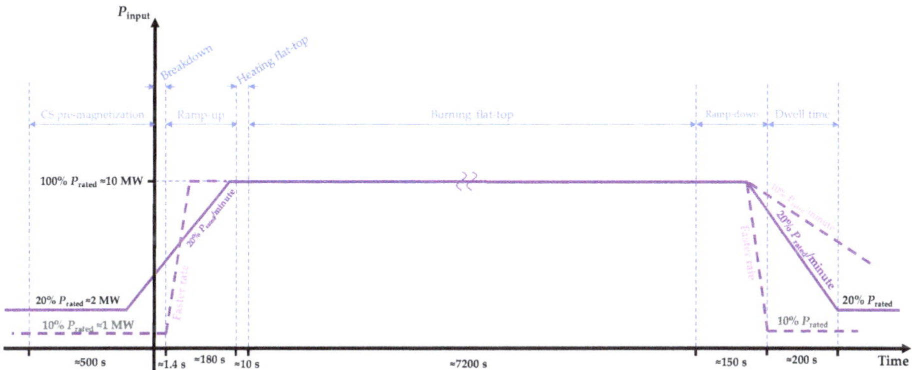

Figure 6. Possible qualitative time evolutions of the power modulation of each of the 18 HCPB OB and IB BB compressors in order to reduce the total power demand. The assumed tradeoff is 20%/minute, a slower rate makes no sense, a faster rate could be detrimental for the components.

In the IHTS there is an intermediate heat exchanger based on "tubes and shell" technology: helium flows into the tubes and molten salt crosses the shell side [13]. In order to mitigate the dynamic behavior of the thermal fusion power between burn flat-top and dwell time on PCS components (particularly on TG and then on the electrical grid), the IHTS is equipped with an ESS. Its task is to store thermal power during burn flat-top phase (removed by the molten salt) and its delivery to the PCS in order to let the TG work almost in steady-state at 80% of the PHTS-rated power, thus avoiding thermo-mechanical cycling issues. Its main electrical loads are the electrical motors moving the circulators present in the loops, whose electrical power was estimated from the thermo-hydraulic design of the IHTS system in terms of mass flow rate and pressure drops. Besides, many other electrical loads regarding all auxiliary systems are still to be evaluated both belonging to the PHTS and the IHTS system. Therefore, they are not included in the present estimations since currently under design.

5.6. HCPB PBS 52: Primary Heat Transfer System (WCLL)

The PHTS based on the WCLL collects all the components and systems devoted to the fusion heat recovery and delivery to the PCS. The main difference with respect to the HCPB architecture is that the heat exchanger is not between PHTS and PCS in the indirect coupling configuration but it receives and stores only the portion of thermal power coming from the first wall of the vessel. This solution aims to reduce the size of the IHTS and its ESS but needs the presence of a second steam generators between the PHTS and the IHTS. However, the same considerations already mentioned for the HCPB hold for the WCLL cooling system about the coupling between the PHTS and PCS as well as the electrical loads to be included in the power flow analysis.

The PHTS consists of two different cooling systems, removing heat from two different parts of the blanket, namely the breeding zone and the tokamak first wall (addressed in PBS 59) and the VV (addressed in PBS 49). Both PHTSs consist of two loops, in order to limit the size of piping and components. The breeding-zone PHTS loops include a steam generator, two pumps, one pressurizer and one hot/cold leg. The first-wall PHTS loops include one heat exchanger, one pump, one pressurizer and one hot/cold leg. Cold/hot ring headers supply/return water to the breeding-zone/first-wall loops. Summing all up, six main coolant pumps are foreseen in the present PHTS WCLL design: four main coolant pumps, two per loop, are present in the cold legs of the breeding-zone PHTS and two main coolant pumps in the first-wall PHTS.

The electrical loads are the electrical motors moving the pumps as well as the pressurizers. Conversely from the motor pumps, which absorb constant electrical power during the whole scenario, the pressurizers are "time-random" electrical loads. They are resistors benches which heat the water up in order to increase the water pressure in pipes whenever the control system sense a pressure reduction. Therefore, they may or may not be switched on during the scenario and not at the same electrical power. In practice, the total power of the pressurizer is 2 MW in standard conditions but when the pressure drops below the minimum threshold the pressurizer is switched on, demanding 10 MW. The very conservative worst-case assumed in these analyses consists in assuming the pressurizers working continuously.

As shown in Figure 6, in the HCPB option, the helium pumps are modulated to operate at reduced power around the dwell time. This could be theoretically useful also for the PHTS-WCLL, but the limited energy saving that could be achieved does not justify the cycling stress produced on the pumps by the frequent power variations.

The WCLL IHTS is conceptually similar to that of the HCPB BB in terms of thermo-hydraulic architecture. First estimations regarding the electrical power absorbed by the electrical motors moving the circulators present on the loops are based on the estimations of mass flow rate and pressure drop in the pipes.

5.7. PBS 60: Remote Maintenance (RM) System

In new-generation tokamaks, Remote Maintenance (RM) is becoming more and more demanding and important with respect to existing facilities in terms of electrical power demand and nuclear safety issues. In fact, in ITER and DEMO the Occupational Radiation Exposure is expected to be twice or three times that of modern nuclear fission plants.

The DEMO RM is mainly divided into two groups:

- In-vessel RM includes all the activities regarding the movement of parts of the vessel or any operation inside it.
- Ex-vessel RM, instead, includes maintenance in and around the tokamak port cells and those in the Maintenance Facility.

The concepts of the safe transportation of extracted parts of the VV are similar to those adopted in ITER. In fact, once extracted from the vessel, both the divertor and the blanket sectors are moved by transport casks through gallery corridors.

Besides, many crane-based systems and lifting systems need to be foreseen in designing the handling of the extracted BB (through the upper port) and divertor (through the lower port) sectors, to move them to the Active Maintenance Facility. Some megawatts of electrical power are expected to be absorbed by this system based on the first estimations on the weight of the parts to be moved.

The total DEMO maintenance and safety needs can be sustained by a RM of about 5 MW. Even though the RM electrical loads belong to the SSEN, parts of them are "decoupled by the plasma operation", as for the remote handling inside the VV. This is a key point in the design of the electrical distribution: the coincidence factor of many electrical loads is zero. However, it is worth mentioning that some RM activities such as the processing/detritiation of some components could be carried out both during and off operations in the Active Maintenance Facility (Building 21 in Figure 3).

5.8. PBS 70: Balance of Plant (HCPB)

Components belonging to PBS 70 are those of the PCS with HCPB-PHTS. This system consists of a classical Hirn Cycle with superheated steam, with steam generator, re-heater, deaerator, condenser, feed water and turbine (connected to the electrical generator for the thermal-electrical energy transformation).

First studies about absorbed electrical power by PCS in case of helium cooled blanket only regard the water pump, based on the requested mass flow rate inside the circuit and the estimated pressure drops. Further investigations will regard ancillary systems serving the PCS, not available at this stage.

It is worth noticing that the net efficiency of the thermal cycle is about 31%, when taking into account the electrical power absorbed by PHTS (in particular, the very high power absorbed by the circulators), IHTS and PCS components. The previous considerations lead to a power estimation of 12 MW for PBS 70.

5.9. PBS 72: Balance of Plant (WCLL)

As for PBS 70, the components belonging to PBS 72 are those of the PCS in the cooling system layout based on the WCLL concept. The same considerations already reported for PBS 70 were proposed, except for slight differences in terms of the steam generator coupling the PCS and the IHTS in case of indirect coupling. It results in a total power absorbed equal to 12 MW. It is worth noticing that first simulations pointed that the system would be able to operate with about 700 MW of gross electrical power and an efficiency equal about to 34%, considering all electrical power needed to be recirculated for PHTS, IHTS and PCS components needs.

5.10. PBS 81: Cryoplant and Cryodistribution

A preliminary design of cryoplant and cryodistribution is under development both in terms of layout of the plant and of estimation of the power and energy absorbed by the mechanical cryopumps. Among the numerous systems to be kept at cryogenic temperatures, it is worth mentioning: the superconducting coils (at 4 K), the coil casings and supporting structures and the thermal shields of VV, cryostat, ports and cryodistribution. Besides, differently from ITER cryoplant, some systems are foreseen to be refrigerated by independent systems, such as cryopumps for the VV and the divertor, NBI cryopumps, pellet launching system and ECRH superconducting magnets. However, the project is currently at a very preliminary stage. A first guess of the total electrical power absorbed by the DEMO cryoplant and cryodistribution (101.8 MW) has been obtained by scaling that designed for ITER project taking into account the volumes of the superconducting magnets and of the other structures and components to be kept at cryogenic temperatures.

5.11. PBS 83: Buildings

The electrical loads included in such estimation regard the lighting systems, the elevator systems and the Heating, Ventilation and Air Conditioning (HVAC) system. Since the design of the DEMO buildings is at a very preliminary stage, a first guess of 54.8 MW is provided by extrapolating the data available from ITER and considering the DEMO preliminary site layout shown in Figure 3.

5.12. PBS 85: Plant Control System

A first guess of 3.6 MW for the power absorbed by the COntrol, Data Access and Communication (CODAC), Central Interlock System (CIS) and Central Safety System (CSS) systems was obtained by estimating the number of cubicles for each subsystem, from those present in ITER.

5.13. PBS 87: Auxiliaries

First preliminary information about the power absorbed by auxiliaries is an extrapolation from ITER's electrical demand for the Component Cooling Water System (CCWS), CHiller Water System (CHWS), sulfur hexafluoride (SF_6) distribution system and compressed air distribution system. This extrapolation results in 90.9 MW.

5.14. Summary of the Total Electrical Power Absorbed by the DEMO SSEN

Since the BoP is still under design also in terms of choice between the two helium-based and water-based cooling system, two parallel and independent designs are under evaluation also in terms of the power electrical system. Figure 7 summarizes the SSEN active and reactive rated powers in each plasma phase for both the two cooling system options. It is important to note the power variations in the HCPB option (Figure 7a) due to the ramp modulation of the compressors motor in the PHTS circuits.

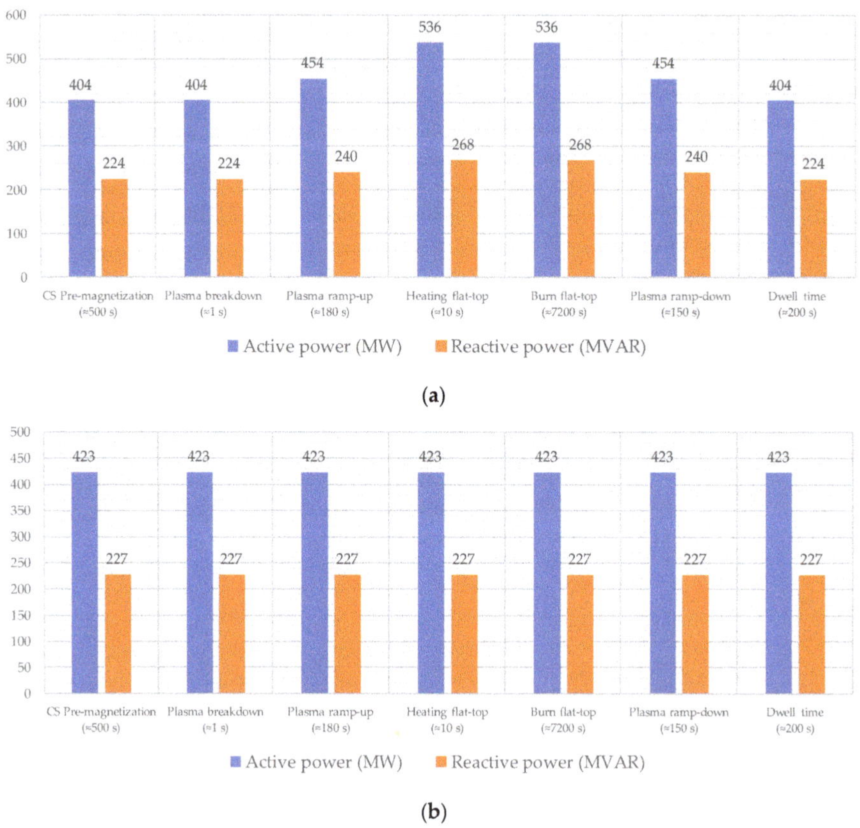

Figure 7. Nominal SSEN active and reactive powers in each plasma phase, considering the two different PHTS-PCS options: (**a**) HCPB and (**b**) WCLL.

6. Preliminary Design and Sizing of the DEMO Electrical Distribution

In compliance with the preliminary scheme in Figure 4 and the lesson learnt from ITER electrical design, four voltage levels are firstly assumed to be used for the DEMO electrical distribution system:

- 22 kV, 6.6 kV and 0.4 kV, in SSEN.
- 22 kV and 66 kV, in PPEN.

According to this assumption and to the aforementioned loads characterization, the HV/MV and MV/LV transformers and the cables for SSEN are sized, by using simulation models for the power flow analysis [19]. The electrical scheme has been designed and implemented in the DIgSILENT PowerFactory software environment. The criteria used for the simulations are:

- Utilization and contemporaneity factors equal to 1 (no redundancy is considered for electrical loads which are supposed to work at their rated power).
- Load power factors assumed to be equal to those identified in the electrical load list.
- Constraint on the nodal voltage variation during operation accepted: <4%.
- Transformers size check in reference of the nominal values.
- Cables size check in reference of the nominal values and to the layout assumptions.

The flow diagram in Figure 8 summarizes the algorithm adopted for the design of the electrical distribution system and for the sizing of the electrical components.

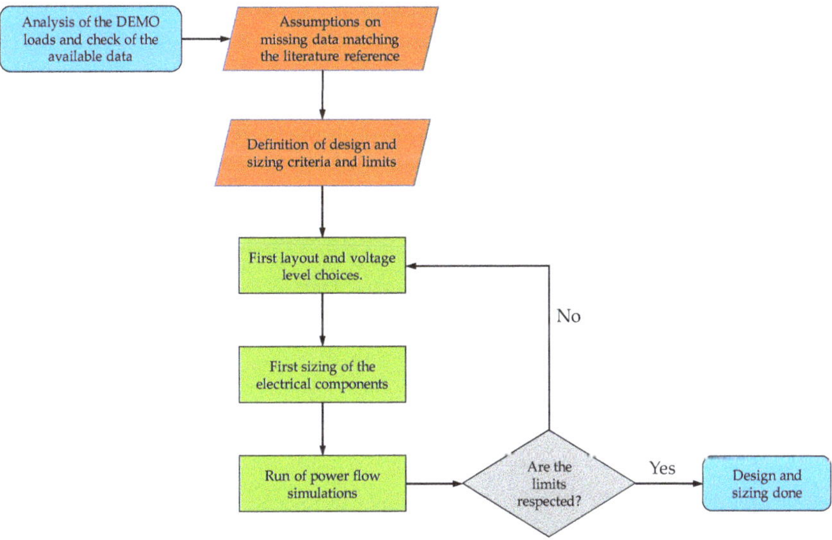

Figure 8. Simplified flow diagram of the algorithm adopted for the preliminary design of the DEMO SSEN.

Simulations were carried out independently for both the two different cooling system configurations (PHTS-HCPB and PHTS-WCLL) and referring only to SSEN loads in each plasma operational phase. As first output of the power flow analysis, the number and the size of the HV/MV transformers are identified according to available data of installed transformed and reliability issues both during normal and off-normal conditions: six three-phases 2-windings transformers with rating power of 150 MVA and voltage ratio equal to 400/22 kV.

As regards the internal distribution for SSEN, the loads have been dispersed following design criteria based on the voltage levels, loads type and distance between the distribution node and the actual location of the loads in the facility estimated on the site layout. The ordinary and the safety loads are distinguished at the 0.4 kV voltage level and the following MV/LV transformers are selected following the same aforementioned criteria:

- Six three-phases 2-windings transformers with rated power of 2.5 MVA and voltage ratio equal to 22/0.4 kV, for the ordinary loads.
- Four three-phases 2-windings transformers with rated power of 1.6 MVA and voltage ratio equal to 22/0.4 kV, for the safety loads.
- Six three-phases 2-windings transformers with rated power of 3.15 MVA and voltage ratio equal to 22/0.4 kV for the supply of some special loads (as those in PBS 22), which are installed separately because of the greater distance from the distribution node with respect to the other loads.

The choice to divide the MV/LV transformers in several groups connected by a bus-tie aims at limiting the short-circuit currents, thus allowing the selection of commercial low voltage circuit breakers. Regarding the loads connected at the 6.6 kV busbars, they are supplied through:

- Five three-phases 2-windings transformers with rated power of 40 MVA and voltage ratio equal to 22/6.6 kV in case of HCPB.

- Five three-phases 2-windings transformers with rated power of 20 MVA and voltage ratio equal to 22/6.6 kV in case of WCLL.

In the 6.6 kV section, the ordinary loads are supplied by 10 three-phases 2-windings transformers with rated power of 40 MVA and voltage ratio equal to 22/6.6 kV. Finally, the cryogenic system is supplied by five three-phases 2-windings transformers with rated power of 40 MVA and voltage ratio equal to 22/6.6 kV. The main data of the selected transformers are reported in Tables 3 and 4, where:

- S_n is the rated power in MVA;
- P_0 is the no-load power in kW;
- P_k is the short circuit power in kW;
- V_k is the short circuit voltage in percentage.

Table 3. Electrical data of the MV/MV and MV/LV transformers.

Transformation Ratio 22/0.4 kV					Transformation Ratio 22/6.6 kV				
S_n [MVA]	P_0 [kW]	P_k [kW]	V_k [%]	I_0 [%]	S_n [MVA]	P_0 [kW]	P_k [kW]	V_k [%]	I_0 [%]
1.6	2.2	13	6	0.9	20	7.7	110	7	0.3
2.5	3.1	19	6	0.6	40	15.1	220	7	0.3
3.15	3.8	22	6	0.5	Winding not present				

Table 4. Electrical data of the HV/MV transformers.

Transformation Ratio 400/22 kV				
S_n [MVA]	P_0 [kW]	P_k [kW]	V_k [%]	I_0 [%]
150	17.1	77.5	13	2

Results obtained by the methodology based on the power flow analysis and the aforementioned sizing criteria allowed also to carry out the cables sizing. It is important to stress that this sizing was possible only by adopting a preliminary but consistent layout as shown in Figure 3. The type of the selected cables and their electrical data are summarized in Table 5, reporting the value of the resistance R' (for the reference and maximum operating temperatures) and reactance X' per unit length. The assumed cable materials and technical characteristics are summarized in Table 6.

Table 5. Electrical data of the SSEN cables.

Rated Voltage [kV]	Section [mm^2]	Rated Current [A]	R' (20 °C) [mΩ/km]	Max R' (90 °C) [mΩ/km]	X' [mΩ/km]
0.4	1 × 240	550	20.1	104	81
0.4	1 × 300	620	64.1	85	79
6.6	1 × 500	760	36.6	51	93
6.6	1 × 630	850	28.3	42	90
6.6	1 × 800	930	22.1	35	86
22	1 × 150	453	124.0	159	128
22	1 × 240	612	75.4	97.8	128
22	3 × 400	813	47.0	62.7	128

Table 6. Materials and technical characteristics assumed for the SSEN cables.

Rated Voltage [kV]	Section [mm^2]	Cores	Conductor	Insulation
0.4	240 300	Single	Stranded flexible annealed bare copper	High-module HEPR rubber, with filler/sheath non-hydroscopic material and special PVC outer sheath
6.6	500 630 800	Single	Copper	Unarmored XLPE, with semi-conductive screens on conductor and insulation
22	150 240 400	Single Triple	Stranded circular compacted copper	XLPE insulation, with conductor screen and insulation screen

From the power flow analysis carried out including the SSEN loads and distribution, the total electrical active and reactive powers requested to the HV point of connection of DEMO to the external electrical grid can be assessed in reference to the different plasma phases. The results are shown in Figure 9 for the HCBP and WCLL cases. This information is the starting point for the sizing and design of the HV switchyard that is object of study and research of the same authors.

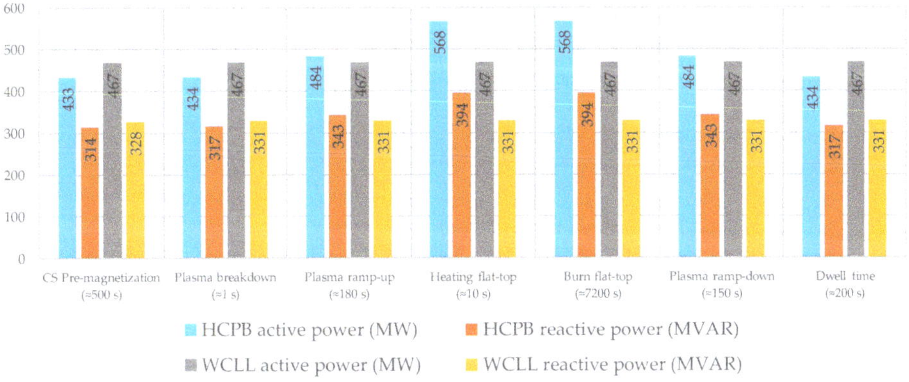

Figure 9. Total steady state electrical powers requested from the grid at the DEMO HV connection point in the two PHTS-PCS cases (HCPB and WCLL).

7. Conclusions

This paper presented the preliminary studies for the feasibility and realization of the demonstrative nuclear fusion power plant DEMO. The study moved from the analysis of the expected DEMO operations and of the electrical loads to sketch a preliminary electric scheme for a part of the internal distribution grid, including the sizing of the main electrical components. This preliminary design was supported by simulation models for the power flow analysis implemented in the DIgSILENT PowerFactory environment.

The first results are focused on the electrical loads analysis and characterization showing a significant variability with respect to the plant configuration, especially because of to the cooling choices and the operational phases. This aspect affects the preliminary design of the electric distribution grid for the SSEN in DEMO. The results show a huge amount of power that would be shared with the HV European grid, pointing out the complexity of the sizing and design of the HV switchyard.

The results shall be continuously refined in the next years following the development of the project and the new experiences in the plasma physics and engineering achieved by ITER or satellite experiments. For instance, the design optimization of the toroidal and blanket segmentation led to a number of sectors being reduced from 18 to 16. The following design choices could lead to a piping redistribution with a limited impact on the total pumping power but there is still room for general design optimizations. Moreover, a relevant breakthrough could consist in the possibility to achieve longer burn phases.

Nevertheless, the electrical design must be continuously updated because its impact is so relevant that could jeopardy the efficiency and the convenience of the plant. In fact, some physics and engineering choices as the cooling configuration and the coil current scenarios could be driven also by electrical constraints or optimizations.

In particular, the following aspects are object of study and research in progress by the same authors:

- Impact of the PPEN power profiles, at least in the reference scenarios.
- Operations and connection of the TG in the direct and indirect coupling.
- Sizing and design of the HV switchyard.

- Technical and legal rules for the connection of DEMO to the HV grid.
- Detailed survey of the critical loads in order to improve the plant safety and reliability and to size the UPSs and the diesel generators with the proper redundancy.
- Location and sizing of the reactive power compensation and filtering systems.

Author Contributions: S.M. collected the data from DEMO PBS experts and from other tokamaks. S.M. and S.P. modeled the distribution network and performed the power flow simulations. S.C. followed the conceptualization of the BoP, of the layout and of the safety classifications. S.P. conceptualized and developed the electrical network. A.L. arranged the electrical load list and the operational phase classification. The methodology was agreed by all the authors. S.M., S.C., M.C.F. and A.L. organized and supervised the project for the involved institutions. The manuscript was edited by all the authors. All authors have read and agreed to the published version of the manuscript.

Funding: This work has been carried out within the framework of the EUROfusion Consortium and has received funding from the Euratom research and training programme 2014-2018 and 2019-2020 under grant agreement No 633053. The views and opinions expressed herein do not necessarily reflect those of the European Commission.

Conflicts of Interest: The authors declare no conflict of interest.

Abbreviations and Acronyms

BB	(Tokamak) Breeding Blanket
BoP	Balance of Plant
CCWS	Component Cooling Water System
CD(R)	Conceptual Design (Review)
CHWS	Chiller Water System
CIS	Central Interlock System
CS	Central Solenoid
CODAC	COntrol, Data Access and Communication
CSS	Central Safety System
DMS	Disruption Mitigation System
ECRH	Electron Cyclotron Resonance Heating
ELM	Edge Localized Mode
ESS	Energy Storage System
EU	European Union
HCPB	Helium Cooled Pebble Bed
H&CD	Heating and Current Drive
HV	High Voltage
HVAC	Heating, Ventilation and Air Conditioning
IB	Inboard Blanket
ICRH	Ion Cyclotron Resonance Heating
IHTS	Intermediate Heat Transfer System
LV	Low Voltage
NBI	Neutral Beam Injector
OB	Outboard Blanket
PBS	Plant Breakdown Structure
PCS	Power Conversion System
PHTS	Primary Heating Transfer System
PPEN	Pulsed Power Electrical Network (classification used in ITER)
RM	Remote Maintenance
SC	Super-Conductor
SSEN	Steady State Electrical Network (classification used in ITER)
TBM	Test Blanket Module
TG	Turbine Generator
VV	Vacuum Vessel
WCLL	Water Cooled Lithium Lead

References

1. Bachmann, C.; Ciattaglia, S.; Cismondi, F.; Federici, G.; Keech, G.; Maviglia, F.; Siccinio, M. Critical design issues in DEMO and solution strategies. *Fusion Eng. Des.* **2019**, *146*, 178–181. [CrossRef]
2. Federici, G.; Bachmann, C.; Barucca, L.; Biel, W.; Boccaccini, L.; Brown, R.; Bustreo, C.; Ciattaglia, S.; Cismondi, F.; Coleman, M.; et al. DEMO design activity in Europe: Progress and updates. *Fusion Eng. Des.* **2018**, *136*, 729–741. [CrossRef]
3. Federici, G.; Bachmann, C.; Barucca, L.; Baylard, C.; Biel, W.; Boccaccini, L.V.; Bustreo, C.; Ciattaglia, S.; Cismondi, F.; Corato, V.; et al. Overview of the DEMO staged design approach in Europe. *Nucl. Fusion* **2019**, *59*, 066013. [CrossRef]
4. EUROfusion. European Research Roadmap to the Realisation of Fusion Energy. 2018. Available online: https://www.euro-fusion.org/eurofusion/roadmap/ (accessed on 15 March 2020).
5. Lampasi, A.; Maffia, G.; Alladio, F.; Boncagni, L.; Causa, F.; Giovannozzi, E.; Grosso, L.A.; Mancuso, A.; Micozzi, P.; Piergotti, V.; et al. Progress of the Plasma Centerpost for the PROTO-SPHERA Spherical Tokamak. *Energies* **2016**, *9*, 508. [CrossRef]
6. ITER Website. Available online: http://www.iter.org/ (accessed on 16 March 2020).
7. Ciattaglia, S.; Federici, G.; Barucca, L.; Stieglitz, R.; Taylor, N. EU DEMO safety and balance of plant design and operating requirements. Issues and possible solutions. *Fusion Eng. Des.* **2019**, *146*, 2184–2188. [CrossRef]
8. Ciattaglia, S.; Federici, G.; Barucca, L.; Lampasi, A.; Minucci, S.; Moscato, I. The European DEMO Fusion Reactor: Design Status and Challenges from Balance of Plant Point of View. In Proceedings of the 17 IEEE International Conference on Environment and Electrical Engineering (EEEIC 2017), Milan, Italy, 6–9 June 2017.
9. Novello, L.; Coletti, A.; Baulaigue, O.; Dumas, N.; Ferro, A.; Gaio, E.; Maistrello, A.; Lampasi, A.; Zito, P.; Matsukawa, P.; et al. Advancement on the Procurement of Power Supply Systems for JT-60SA. In Proceedings of the IEEE 25th Symposium on Fusion Engineering (SOFE), Austin, TX, USA, 31 May–4 June 2015.
10. Lampasi, A.; Zito, P.; Starace, F.; Costa, P.; Maffia, G.; Minucci, S.; Gaio, E.; Toigo, V.; Zanotto, L.; Ciattaglia, S. The DTT device: Power supplies and electrical distribution system. *Fusion Eng. Des.* **2017**, *122*, 356–364. [CrossRef]
11. Morris, J.; Kovari, M. Time-dependent power requirements for pulsed fusion reactors in systems codes. *Fusion Eng. Des.* **2017**, *124*, 1203–1206. [CrossRef]
12. Lampasi, A.; Minucci, S. Survey of Electric Power Supplies Used in Nuclear Fusion Experiments. In Proceedings of the 17 IEEE International Conference on Environment and Electrical Engineering (EEEIC 2017), Milan, Italy, 6–9 June 2017.
13. Barucca, L.; Ciattaglia, S.; Chantant, M.; Del Nevo, A.; Hering, W.; Martelli, E.; Moscato, I. Status of EU DEMO heat transport and power conversion systems. *Fusion Eng. Des.* **2018**, *136*, 1557–1566. [CrossRef]
14. Moscato, I.; Barucca, L.; Ciattaglia, S.; Di Maio, P.; Federici, G. Preliminary design of EU DEMO helium-cooled breeding blanket primary heat transfer system. *Fusion Eng. Des.* **2018**, *136*, 1567–1571. [CrossRef]
15. Ferro, A.; Lunardon, F.; Ciattaglia, S.; Gaio, E. The reactive power demand in DEMO: Estimations and study of mitigation via a novel design approach for base converters. *Fusion Eng. Des.* **2019**, *146*, 2687–2691. [CrossRef]
16. Gliss, C.; Ciattaglia, S.; Korn, W.; Moscato, I. Initial layout of DEMO buildings and configuration of the main plant systems. *Fusion Eng. Des.* **2018**, *136*, 534–539. [CrossRef]
17. Falvo, M.C. Studies on the generator operation assuming power profiles corresponding to the PHTS-PCS indirect and direct cycle, v1.0, 20 December 2019, EUROfusion Report IDM: EFDA_D_2NUUME. Available online: https://idm.euro-fusion.org/ (accessed on 15 March 2020).
18. Del Nevo, A. Direct Coupling of WCLL BB PHTS to PCS feasibility study, v1.0, 14 February 2019, EUROfusion Report IDM: EFDA_D_2MN55V. Available online: https://idm.euro-fusion.org/ (accessed on 15 March 2020).
19. Duncan Glover, J.; Sarma, M.S.; Overbye, T.J. *Power System Analysis and Design*, 5th ed.; Cl-Engineering: Pacific Cove, CA, USA, 2011.

Article

Techno-Economic Evaluation of Interconnected Nuclear-Renewable Micro Hybrid Energy Systems with Combined Heat and Power

Hossam A. Gabbar [1,2,*], Muhammad R. Abdussami [1] and Md. Ibrahim Adham [1]

[1] Faculty of Energy Systems and Nuclear Science, Ontario Tech University (UOIT), Oshawa, ON L1G 0C5, Canada; mdrafiul.abdussami@ontariotechu.net (M.R.A.); mdibrahim.adham@ontariotechu.net (M.I.A.)

[2] Faculty of Engineering and Applied Science, Ontario Tech University (UOIT), Oshawa, ON L1G 0C5, Canada

* Correspondence: hossam.gaber@uoit.ca

Received: 7 March 2020; Accepted: 31 March 2020; Published: 2 April 2020

Abstract: Renewable energy sources (RESs) play an indispensable role in sustainable advancement by reducing greenhouse gas (GHG) emissions. Nevertheless, due to the shortcomings of RESs, an energy mix with RESs is required to support the baseload and to avoid the effects of RES variability. Fossil fuel-based thermal generators (FFTGs), like diesel generators, have been used with RESs to support the baseload. However, using FFTGs with RESs is not a good option to reduce GHG emissions. Hence, the small-scale nuclear power plant (NPPs), such as the micro-modular reactor (MMR), have become a modern alternative to FFTGs. In this paper, the authors have investigated five different hybrid energy systems (HES) with combined heat and power (CHP), named 'conventional small-scale fossil fuel-based thermal energy system,' 'small-scale stand-alone RESs-based energy system,' 'conventional small-scale fossil fuel-based thermal and RESs-based HES,' 'small-scale stand-alone nuclear energy system,' and 'nuclear-renewable micro hybrid energy system (N-R MHES),' respectively, in terms of net present cost (NPC), cost of energy (COE), and GHG emissions. A sensitivity analysis was also conducted to identify the impact of the different variables on the systems. The results reveal that the N-R MHES could be the most suitable scheme for decarbonization and sustainable energy solutions.

Keywords: nuclear power plant; renewable energy; hybrid energy system; combined heat and power

---

1. Introduction

According to the 2030 Agenda for Sustainable Development, seventeen Sustainable Development Goals (SDGs) are set for action by both developed and developing countries. Among the seventeen SDGs, one of the most crucial goals is "Affordable and Clean Energy." The objective of the goal "Affordable and Clean Energy" is to ensure not only affordable, resilient, and modern forms of energy, but also sustainable and carbon-free electricity for the planet [1].

Electricity is a fundamental requirement for global advancement and economic growth. The demand for electricity is growing proportionally to the population and economic development. Currently, the world is undergoing two challenges for managing the high demand for electricity; one problem is how to support the high demand for electricity without exploiting finite energy resources (mainly fossil fuels), and another challenge is how to produce electricity without affecting the environment [2].

Currently, most of the electric energy is produced from conventional sources like coal, gas, and oil. The production of electricity by using these resources tends to raise the emission of greenhouse gases (GHGs) in the atmosphere. Research is ongoing to reduce the impact of using traditional sources for electricity generation in the environment. For instance, one such initiative is the implementation of carbon capture and storage (CCS) policies. In this process, the waste carbon dioxide from the power plant can be captured and transported to a storage site for disposal so that it cannot disperse in the environment [3].

In recent times, the world is looking for sustainable energy sources that will be used to meet today's demand without putting them into danger for future usages. Consequently, renewable energy sources (RESs), such as solar, wind, geothermal, hydropower, and ocean energy, are being recognized as sustainable sources for electric energy production [4]. RESs are intermittent, and electricity cannot be stored economically for an extended period, so some other sources of energy that can provide a back-up for the RESs during their unavailability period and act as a base load or critical load supplier are needed.

The research and development plan for dynamic modeling and simulation of large scale nuclear-renewable hybrid energy system (N-R HES) is addressed in [5]. The N-R HES is interpreted as a combination of a nuclear reactor, a turbine for electricity generation from thermal energy, at least a RES, and a product produced from the N-R HES by an industrial process. In the report, the authors categorized the N-R HES into three types: tightly coupled HES, thermally coupled HES, and loosely coupled electricity-only HES. The potential vital benefits of N-R HES include GHG-free electricity, a resilient electric grid, and low COE. The authors have also regarded the integration of SMR and RES for future work.

A combination of renewable energy generation, nuclear reactors, and industrial processes, including the versatility of grid and making the best possible use of investment, have been explored in [6]. Six features of interconnection have been identified here—electrical, thermal, chemical, hydrogen, mechanical, and information. This study concluded that the integration of nuclear and renewable energy could be a potential solution for a long-term and ample amount of power and heat supply that is free from sudden price changes like fossil-fuels. This document also pointed out that the nuclear-renewable hybrid system can supply load-following power, and excess energy can be used for the production of secondary energy-intensive products. Nevertheless, system analysis, technical advancement, and optimization are required to implement this hybrid system in practice.

Currently, the International Atomic Energy Agency (IAEA) has published a technical document of nuclear-renewable integration. The report addresses the national strategies on nuclear and RESs, opportunities and challenges of nuclear-renewable integration, and the role of the small-scale reactors in nuclear-renewable hybridization [7].

A comparison of three scenarios, namely a nuclear power plant, a combination of a nuclear plant and a wind facility, and a mixture of nuclear and wind energy sources with a hydrogen generation facility, have been done in [8]. This study reported that with optimization, the nuclear-wind system with a hydrogen production facility could be an economically viable option in the future. Sensitivity analysis has also been carried out to realize the impact of the energy market, depreciation rate, discount rate, and time horizon in terms of internal rate of return (IRR), levelized cost of energy (LCOE), net present value (NPV), and payback period.

In [9], the author highlighted the critical challenges of nuclear-renewable integration, such as integration values, regulatory, financial, technological, plant testing, and plant operation. The author suggested that the information linkage-based nuclear-renewable coupling would be able to overcome the complexities of the integration process.

Three scenarios of N-R HES to supply thermal energy from the system have been examined in [10]. The first arrangement includes a nuclear reactor, thermal power cycle, wind power plant, and electric boiler; the second scenario comprises a nuclear reactor, thermal power cycle, wind power plant, and electric thermal storage; the third configuration is a combination of a nuclear reactor, thermal power cycle, wind power plant, electric boiler, and thermal storage. The electric thermal storage stores thermal energy that is generated by electricity. The financial performance analysis tells us that the third arrangement has the lowest NPV, lowest IRR, and highest TCI. The analysis results are evident because the thermal power supply is elevated significantly by introducing the electric boiler and the thermal storage simultaneously in the third scenario. The result is also evident since the authors assumed that the cost of heat generated from the nuclear reactor is less than the price of heat generated from gas (electric boiler) [10].

The Hybrid Optimization Model for Electric Renewable (HOMER) software has been used for predicting the viability of a HES in [11]. The HES consists of photovoltaic (PV) cells, diesel generator, and battery bank. Paper [11] summarized that the PV-diesel-battery system has one-third fuel-savings compared to the diesel-only system. It is also concluded that hydrogen technologies can technically replace the diesel-battery system. The PV-diesel-battery configuration will also reduce emissions and increase the penetration of renewable energy. Besides, the authors found that the implementation of 20% renewable energy would reduce a massive amount of CO_2 compared to the present value of emission [11].

However, small scale nuclear reactors like the micro modular reactor (MMR) are gaining attention currently for their small size, affordability, security, reliability, and innovativeness. Several companies are now working on these types of reactors. For example, U-Battery, an MMR manufacturer, is expecting to demonstrate this type of reactor by 2026 [12]. Some other very small reactors are being developed, such as the eVinci^{TM} micro reactor with combined heat and power (CHP) rated from 200 kWe to 5 MWe [13]. Consequently, these types of micro reactors can replace the FFTGs (diesel generators), which are now being employed with RESs either as the backup power supply or as the main electric power- generating source.

In this paper, the planning and optimization of different HES models have been developed and evaluated based on multiple techno-economic key performance indicators (KPIs). The grid-connected as well as off-grid mode have been analyzed in this paper. Five different scenarios with CHP have been developed and assessed using the HOMER software. Comparison among all situations has been made based on three KPIs - NPC, COE, and GHG emission. This document is divided as follows: the detailed nuclear-renewable hybrid energy system is discussed in Section 2. System configurations are addressed in Section 3. Section 4 covers the design considerations that are made for system evaluation. Part 5 illustrates the system simulations. The simulation results, based on selected KPIs, are compiled in Section 6. The discussion is presented in Section 7.

2. Nuclear-Renewable Micro Energy System

N-R HES is a collective network of different RESs, nuclear reactors, energy storage systems (ESSs), power electronic devices, and various energy users (e.g., electric, thermal, and hydrogen). Since no fossil fuel is combusted in the N-R HES, it is the cleanest HES with virtually zero GHG emissions. The N-R HES utilizes a substantial amount of waste heat energy from the thermal generator (e.g., nuclear reactor, geothermal energy, concentrated solar power, and biomass) to generate different products. Several advanced control algorithms are used in N-R HES to ensure the security and reliable performance of the system. Based on the size of the hybrid system, the coupling scheme can be categorized into two types [14], as discussed below.

2.1. Large-scale Coupling

In large-scale coupling, a traditional large-scale nuclear power plant (NPP) is collocated with regionally available RESs. The coupling may occur at either the electrical, thermal, or electrical-thermal levels. For example, the International Atomic Energy Agency (IAEA) recommends that a conventional large-scale NPP must have a bare area—called "exclusion zone"—around the NPP for safety purposes. A part of this exclusion zone can be used to install wind turbines to extract wind energy and integrate it with the NPP generation or electric grid [15]. Since the exclusion zone is typically a large empty area, it might be a favorable space of achieving high wind speeds, implying a high wind energy potential. Moreover, e PV panels can be fitted on the different facilities of the NPP to harness solar power.

The large-scale coupling also includes the mobile microgrid (MM) with traditional NPP for various purposes. An MM is a HES consisting of different RESs, such as wind and PV, with intelligent remote control capability for resilient off-grid power supply [16]. An MM can also be linked with large-scale NPP for emergency cases to support the essential electric system, Class I power supply, of the NPP. Class I, associated with a battery bank, is the most sensitive power class in NPP, and it can never be interrupted. Hence, a battery fast charging mechanism, powered by MM, can solve the drawbacks of the battery bank and ensure the safety of NPP. In summary, the integration of MM with traditional NPP works for both objectives – load demand fulfillment and NPP emergency back-up support [17].

2.2. Modular-scale Coupling

In modular-scale coupling, a small-scale reactor, called small modular reactor (SMR) or micro modular reactor (MMR), is conjoined with RESs at a site where the RESs are mostly available. There are notable advantages of MMR, explained in the later section, over the SMR. The MMR is moveable and modular in size; the MMR is towed into a suitable location to combine with RESs in modular-scale coupling schemes. The HES that uses a modular-scale coupling method is called nuclear-renewable micro hybrid energy system (N-R MHES). The N-R MHES has immense applications in remote community service, transportation electrification, distant oil and gas mining facilities, and remote chemical industries. The grid-connected N-R MHES also provides an excellent energy solution to the medium-level electricity demand with the lowest NPC and COE [14].

3. System Configuration

A case study has been conducted to assess some specific KPIs—net present cost (NPC), cost of energy (COE), and greenhouse gas (GHG) emissions—of an N-R MHES. The grid-level coupling method is reflected in the study. The case study includes the system equipment with some conservative assumptions that are listed with each component in the subsections. The load data are obtained from the National Aeronautics and Space Administration (NASA) Surface Meteorology and Solar Energy database [18]. The data are synthesized in a MATLAB simulator, and the Hybrid Optimization of Multiple Energy Resources (HOMER) Pro software is used for system modeling, simulations, and KPI analysis. In the study, the project lifetime is assumed as a whole 30 years – starting in 2007 and ending in 2036. The prices of the system component are pragmatic compared to the present market value. The RESs data, such as solar radiation, wind speed, and water flow of the hydro dam, are assumed to be the same during the complete project lifetime. However, some important system parameters, such as load demand and equipment cost, change in each year of project life. To address the effect of these variations on the whole system, an additional study is conducted by "Multiyear Analysis" module in the HOMER Pro software. The details of individual system equipment are demonstrated in the next few subsections. The project life, inflation rate, and discount rate are considered as 30 years, 2%, and 8%, respectively. The costs of all system equipment, mentioned in the specification table in this study, are the first-year cost of the whole project life.

The HOMER software uses the inflation rate and discount rate to calculate the actual discount rate. The actual discount rate is used to determine the discount factors and annualized costs. The inflation rate, discount rate, and actual discount rate are mathematically related as follows [19]:

$$r = \frac{d - i}{1 + i} \tag{1}$$

where r, d, and i represent actual discount rate, discount rate, and inflation rate, respectively.

3.1. Electric Load

The electric load data has been collected for the year of 2018 from several facilities of Ontario Tech University (UOIT, Oshawa, ON, Canada). These data are deemed as input electric load data for the first year of the project life. The data are collected from UOIT since it has a significant variation in the load profile of the UOIT facilities. As a micro HES is considered for this analysis, the actual load profile is scaled-down to a comparable size to emulate the load profile of a small-scale energy system.

The scaled-down data also reduces the simulation time. Nevertheless, the scaled-down process does not change the load characteristics.

The load profile that is collected from UOIT is designed and named "Electric Load #1" in the HOMER Pro software interface. Another electric load is modeled in HOMER Pro, entitled "Electric Load #2," to create "diversity" in the system load profile. The peak occurs for "Electric Load #1" at noon, whereas the peak occurs for "Electric Load #2" at evening. Thus, the diversity is created. A deferrable electrical load is also designed in the HOMER Pro software. The value of average energy demand, storage capacity, peak power demand, and minimum load ration of the deferrable load are 12 kWh/day, 48 kWh, 8 kW, and 80%, respectively. The "Electric Load #1" and "Electric Load #2" are given the highest priority for demand fulfillment in the system and considered as the primary load in simulation; whereas, the deferrable electric load gets less priority. In the simulation, the resources always supply electricity to the primary load first, then go for the deferrable load. Figure 1 illustrates the total electric load profile - summation of Electric Load #1, Electric Load #2, and deferrable load - of the system. Table 1 represents the details of system electric load.

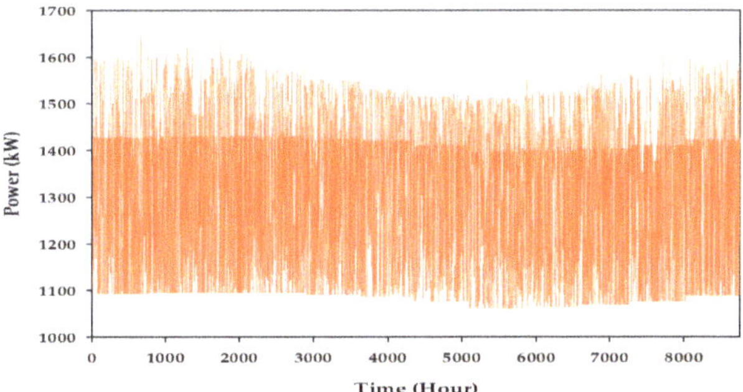

Figure 1. Total electrical load profile (first year of project lifetime).

Table 1. Specifications of the primary electric load.

Parameter	Value	
	Electric Load #1	Electric Load #2
Average Energy Demand (kWh/day)	12,000	500
Average Power Demand (kW)	500	20.83
Peak Power Demand (kW)	857.17	61.83
Load Factor (%)	58	34

The yearly total electrical load profiles are depicted in Figure 1. The characteristics of the electrical loads are summarized in Table 1.

3.2. Thermal Load

Since the actual thermal load data is not available for the project, a standard thermal load profile is modeled in this study from HOMER Pro software library. Two types (commercial load and community load) of thermal loads are designed to introduce variety into the N-R MHES. The thermal loads are compatible and comparable to the typical hybrid energy system.

The yearly total thermal load profiles are depicted in Figure 2. The peak demand occurs at noon and evening for Thermal Load #1 and Thermal Load #2, respectively. The characteristics of thermal loads are summarized in Table 2.

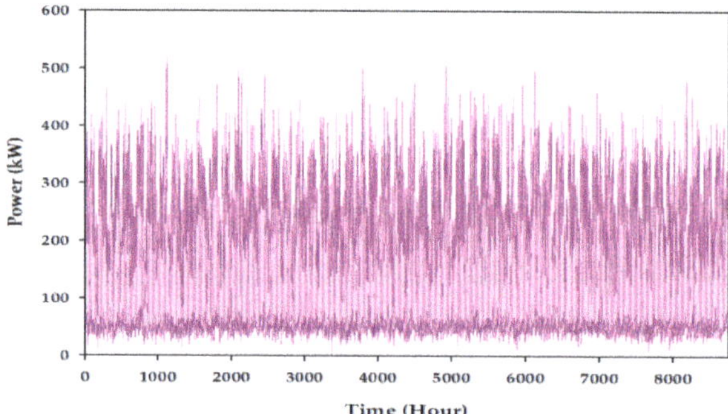

Figure 2. Total thermal load profile (first year of project lifetime).

Table 2. Specifications of the thermal load.

Parameter	Value	
	Thermal Load #1	**Thermal Load #2**
Average Thermal Energy Demand (kWh/day)	3500	200
Average Thermal Power Demand (kW)	143.83	8.33
Peak Thermal Power Demand (kW)	502.55	24.73
Load Factor (%)	29	34

3.3. Diesel Generator

Though the FFTG is not a compulsory part of N-R MHES, a Diesel genset is studied here to investigate and compare the KPIs between conventional fossil fuel-based thermal hybrid energy systems and N-R MHES. The traditional hybrid energy system combines FFTGs, RESs, energy storage, and power electronic components. In this project, a 1 MW diesel generator is compared with a 1 MW MMR, side by side, to obtain the most feasible system in terms of economic and environmental viewpoints. The heat recovery ratio and diesel price are assumed 40% and 0.79 UDS/Liter [20], respectively in this study. The economic parameters of a typical diesel generator, used in the simulation, is listed in Table 3. The HOMER Pro software automatically calculates the total emission by diesel generator while the Diesel genest produces electricity.

3.4. Solar Power

Solar photovoltaic (PV) technology is used in this project to convert solar radiation into useful electricity. The solar radiation data of UOIT have been collected from NASA Surface Meteorology and Solar Energy database. The solar radiation data are assumed to be the same throughout the 30 years of the project lifetime.

The output of the PV array is calculated as follows [21]:

$$P_{PV} = Y_{PV} f_{PV} \left(\frac{G_T}{G_{T,STC}} \right) \left[1 + \alpha_p (T_c - T_{c,STC}) \right] \tag{2}$$

where Y_{PV}, f_{PV}, G_T, $G_{T,STC}$, α_p, T_c, and $T_{c,STC}$ denote the rated capacity of the PV array under standard test conditions (kW), PV derating factor (%), solar radiation incident on PV in the current time step (kW/m^2), solar incident radiation at standard test condition (1 kW/m^2), temperature coefficient (%/°C), PV cell temperature in the current time step (°C), and PV cell temperature under standard test conditions (25 °C), respectively.

The manufacturers set the PV power rating at a specific condition called "Standard Test Conditions (STC)." In STC, the solar radiation is assumed to be 1 kW/m^2, the cell temperature is taken as 25 °C, and no wind flow is considered around the PV cell. However, in practical cases, the full-sun cell temperature is always greater than 25 °C, and there is a significant wind flow encompassing the PV system. The derating factor is also regarded to model the PV cell explicitly since it is related to the panel losses, wiring losses, and aging. The derating factor is a scaling factor to match the PV rating of the manufacturers with the real-world conditions.

If the temperature effect is ignored to design the PV panel, Equation (3) can be simplified as follows:

$$P_{PV} = Y_{PV} f_{PV} \left(\frac{G_T}{G_{T,STC}} \right) \tag{3}$$

The yearly solar irradiance data are represented in Figure 3. A flat panel 400 kW PV array, having an operating temperature 25 °C, is used here. The PV array consists of 400 solar cell units; each unit is rated at 1 kW. The details specifications of the solar PV cell are summarized in Table 3.

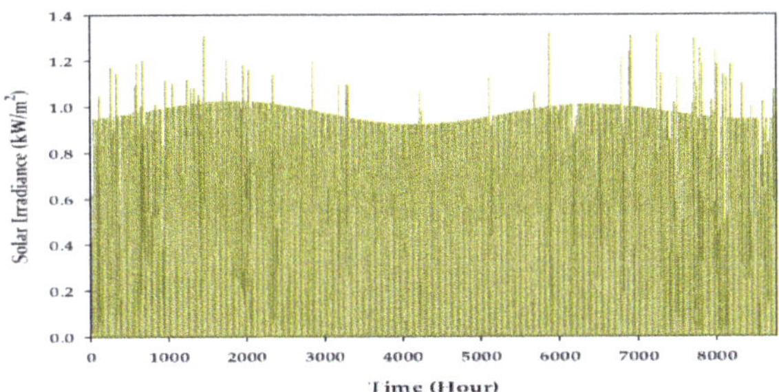

Figure 3. Solar irradiance.

3.5. Wind Power

Two (2) wind turbine units, each rated as 330 kW, are used in this N-R MHES. The hub height and rotor diameter of the turbine are 55 m and 33 m, respectively. It has been assumed that the wind speed is the same for the entire project life. The details of the wind turbine are presented in Table 3.

Three steps are maintained to calculate the output of the wind turbine in HOMER [22]. The first step is to estimate wind speed at the hub height of the wind turbine. The wind speed at the hub height is calculated by Equation (4):

$$V_{HUB} = V_A \frac{\ln\left(\frac{H_{HUB}}{H_0}\right)}{\ln\left(\frac{H_A}{H_0}\right)} \tag{4}$$

where V_{HUB}, V_A, H_{HUB}, H_0, and H_A denote wind speed at the hub height of the wind turbine (m/s), Wind speed at the anemometer height (m/s), Hub height of the wind turbine (m), Surface roughness length (m), and Anemometer height (m), respectively.

The second step is to calculate the turbine power at standard air density; this is done with the help of the turbine power curve given by the manufacturer. The wind power output at standard air density is estimated from the corresponding wind speed value, calculated in the previous step.

The third step deals with the application of density correction. The following equation is used for density correction to calculate the exact wind power extracted by the wind turbine:

$$P_{ACTUAL} = \left(\frac{\rho}{\rho_0} \right) P_{STP} \tag{5}$$

where P_{ACTUAL}, ρ, ρ_0, and P_{STP} represent actual power output of the wind turbine (kW), actual air density (kg/m^3), air density at standard temperature and pressure (kg/m^3), and power output of the wind turbine at standard temperature and pressure (kW), respectively. Figure 4 represents the wind speed profile at UOIT for one year.

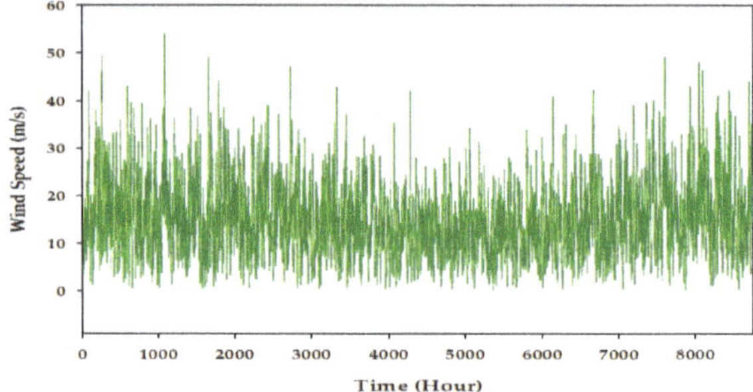

Figure 4. Wind speed.

3.6. Hydro Power

A run-of-river hydroelectric plant is considered in this study. The average yearly (2018) mass flow rates of Lake Ontario, Canada are collected and shown in Figure 5. The economic specifications of the hydro plant and hydro turbine are summarized in Table 3.

The nominal power of a hydropower plant can be calculated from the following equation [23]:

$$P_{HYDRO} = H_A \times \rho_{WATER} \times g \times Q_{TURBINE} \times \eta_{TURBINE} \tag{6}$$

where H_A, ρ_{WATER}, g, $Q_{TURBINE}$, and $\eta_{TURBINE}$ denote available water head (m), water density ($\frac{kg}{m^3}$), gravitational constant ($\frac{m}{s^2}$), mass flow rate ($\frac{m^3}{s}$), and hydro turbine efficiency (%), respectively.

The mass flow rate or hydro turbine flow rate is the amount of water that flows through the hydro turbine. In this study, the mass flow rate is calculated using Equation (7) [23]:

$$Q_{TURBINE} = \begin{cases} 0, Q_{AVAILABLE} < Q_{MINIMUM} \\ Q_{AVAILABLE}, \ Q_{MINIMUM} \leq Q_{AVAILABLE} \leq Q_{MAXIMUM} \\ Q_{MAX}, \ Q_{AVAILABLE} > Q_{MAXIMUM} \end{cases} \tag{7}$$

where $Q_{AVAILABLE}$, $Q_{MINIMUM}$, and $Q_{MAXIMUM}$ represent available mass flow rate to hydro turbine ($\frac{m^3}{s}$), available minimum mass flow rate to hydro turbine ($\frac{m^3}{s}$), and available maximum mass flow rate to hydro turbine ($\frac{m^3}{s}$), respectively.

The available water head, design flow rate, minimum flow rate, and maximum flow rate of the hydroelectric plant are considered as 25 m, 500 L/s, 50%, and 150%, respectively, in this case study.

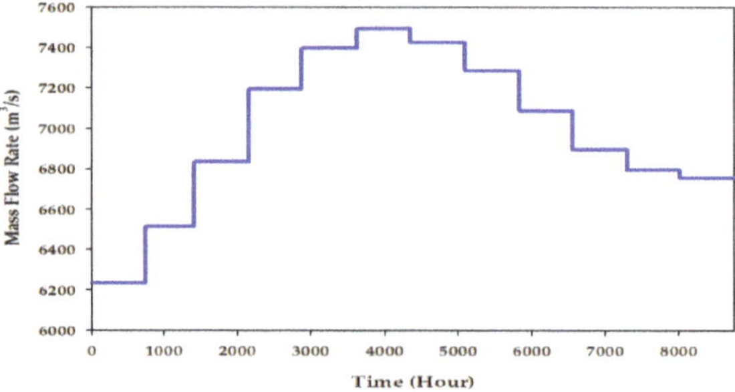

Figure 5. Mass flow rate.

Table 3. Specifications of some system equipment.

Parameters	System Equipment			
	Solar PV	**Wind Turbine**	**Hydro Generator**	**Diesel Genset**
Nominal Capacity (kW)	1	330	98.1	1000
Capital Cost (USD)	640	40,000	459,845	300,000
Replacement Cost (USD)	640	30,000	229,923	200,000
Lifetime (Years)	30	20	25	2.5
Operation and Maintenance Cost (USD/year)	0	100	13,795	87,600

3.7. Nuclear Power Plant

Due to the high initial cost and necessity of a large installation site of the conventional NPP, a micro modular reactor (MMR) is a promising solution for reducing the capital cost and eliminating the need for a massive installation area. MMRs have some benefits over conventional NPPs; MMR takes a shorter time to construct, design of MMR is simple and flexible, and MMR is appropriate for small-scale electricity generation systems [24].

According to the IAEA, the NPP rated under 300 MWe can be interpreted as 'small' NPP. IAEA defines the NPP as 'medium' if the reactors have the power rating up to 700 MWe. The 'small' and 'medium' NPP collectively have been called 'small and medium reactor (SMR),' but commonly, they are termed as 'small modular reactor (SMR).' One subclass of SMR is 'very small reactors (vSMRs)' that are rated under 15 MWe, and are very competent for remote communities [25].

The MMR, a generation-IV reactor, is a small-scale reactor, characterized by a power rating in between 1 MWe to 20 MWe. The MMR provides a safe, emission-free, and cost-efficient energy source for both on-grid and off-grid applications. Due to its modular nature, small footprint, in-factory construction, cogeneration capability, high-level safety measures, insignificant installation area, design simplicity, scalability, and transportability, the MMR has a high impact on energy system modeling consideration. The MMR could be an outstanding solution for large-scale generating station back-up power, remote industries, and transportation electrification [26]. A feasibility assessment of the MMR for military ground application has been conducted in [27]. Based on inherent risk, strategical complication, and cost associated with using diesel generators in a forward operating base (FOB), the authors found that the MMR is better than diesel generator for longer-term operation.

The TRISO fuel, a proven technology, is used in the MMR. TRISO is a uranium fuel coated with three layers; the uranium center is covered with a carbon layer, followed by silicon carbide, followed by an outer carbon layer. This extraordinary design ensures the utmost safety of the fuel under extreme conditions. The TRISO fuel can sustain up to 1800 °C, which is 200 °C hotter than the accident condition. The MMR is a gas-cooled reactor where helium is circulated in a primary circuit, and nitrogen is used in the secondary circuit. Since helium is an inert gas, clean, free of fission product, and does not react chemically with reactor or fuel, helium provides numerous benefits to the MMR. The fuel enrichment is around 9–12% for MMR. The molten salt is used in MMR as thermal storage to supply electricity and process heat. The power density, outlet temperature, and the process heat temperature of MMR are 1.24 W/cm^3, 630 °C, and 750 °C, respectively. The active system is not required to extract heat in the MMR-based plant. Moreover, off-site service, such as electric power, is also not needed for the safe operation of this type of reactor [28].

A 1 MWe MMR is considered in this research. It always supplies a constant 1 MWe to the load. Since the MMR has the CHP capability, 40% of the waste heat of the MMR can be utilized to support the thermal load of the system. There are no GHG emissions to produce electricity from MMR and there is no GHG emissions during operation. However, if fossil fuel is burnt in the construction of NPP, or if the fossil fuel is used in mining or refining uranium ore, GHG will be produced; but, this types of

GHG emissions are not considered in this study. A detail specification of MMR is listed in Table 4. The MMR can operate without refueling up to 5 years [29]. However, the spent fuel transportation cost is not considered in this study.

Table 4. Specifications of the MMR.

Parameter	Value
Fuel Type	Uranium
Nominal Capacity (kW)	1000
Capital Cost (USD)	11,250,000
Refurbishment Cost (USD)	2,300,000
Operation and Maintenance Cost (USD/lifetime)	1,510,000 [30]
Lifetime (Years)	60 [29]
Efficiency (%)	40 [30]
Heat Recovery Ratio (%)	40
Fuel Price (USD/kg)	1390 [31]

The capital cost of an exact 1 MW*$_e$ MMR is also not available. Therefore, the capital cost of a 1 MW$_e$ MMR is calculated as 11.25 million USD by linear estimation from the capital cost of a 2 × 10 MW**$_t$ (2 units, each having a power rating of 10 MW$_t$) [30]. The refurbishment cost of the plant is approximated as 2.3 million USD [31]. (*MW$_e$ = megawatt electric, **MW$_t$ = megawatt thermal).

3.8. Electric Grid

The studied N-R MHES is analyzed for the grid-connected mode of operation. An electric grid, with the "net metering" provision, is considered here. The purchase and the sale capacity of the grid are determined and optimized by the HOMER software. The grid capacities (purchase and sale) are optimized in such a way that the availability of electric power is always guaranteed for all cases, discussed in Sections 5.1–5.5. The optimized grid capacities confirm that if the energy resources fail to supply the electricity demand, the grid is competent in providing the required electric power to the load. The measured purchase and sale capacity of the electric grid are 0.9 MW and 1000 MW, respectively. The purchase capacity is the maximum amount of power that can be consumed from the grid at a time. The sale capacity is the maximum power that can be sold to the grid by "net metering "at any time.

The electric energy that is being supplied to the grid is typically generated by burning fossil fuel. In this study, natural gas is considered as the fuel to generate the grid electricity. As the research also focuses on the environmental consequences of the N-R MHES, the grid emission is also considered. Though natural gas is a clean form of fuel, it still releases different types of gases to the atmosphere. The amount of emission from natural gas is listed in Table 5.

Table 5. Grid emissions [32].

Pollutant	Emission Factor (g/kWh)
Carbon Dioxide (g/kWh)	181.1
Carbon Monoxide (g/kWh)	0.062
Particulate Matter (g/kWh)	0.0108
Sulfur Dioxide (g/kWh)	0.00155
Nitrogen Oxides (g/kWh)	0.1424

Furthermore, the energy rate is not the same for all day long; it varies with time and types of load, such as industrial load, community load, residential load, and commercial load. For example, the electricity price is not equal in the morning, and at the mid of the day. Even the cost of electricity in summer and winter also differs. Similarly, the electricity charge is different for household loads and industries. By studying all these perspectives, different electricity prices are discussed in this study for different times of the day and the year. The energy rates of the electric grid are summarized in Table 6.

Table 6. Energy rates of the electric grid [33].

Schedule		Types of Energy Rate	Electricity Price (USD/kWh)	Sellback Price (USD/kWh)
September-April	Weekday	Off-peak (19.00–07.00)	0.065	0.025
		Mid-peak (11.00–17.00)	0.094	
		On-peak (07.00–11.00 & 17.00–19.00)	0.134	
	Weekend	Off-peak (full day)	0.065	0.025
May-August	Weekday	Off-peak (19.00–07.00)	0.065	0.025
		Mid-peak (07.00–11.00 & 17.00–19.00)	0.094	
		On-peak (11.00–17.00)	0.134	
	Weekend	Off-peak (full day)	0.065	0.025

3.9. Energy Storage

A lithium-ion battery bank, comprising 500 battery cells (each has nominal capacity of 1 kWh), is regarded as the energy storage component in this project. Both the capital and replacement cost of a single cell are considered as 180 USD. The battery lifetime, SOC_{min}, SOC_{min} are 15 years, 100%, and 30%, respectively, for this case.

3.10. Boiler

The boiler produces thermal power by heating fluids inside the boiler vessel to serve the thermal load. Though the generators, considered in this project, have CHP capability, the boiler is kept inside the scenarios to support the thermal load for severe cases. If the extracted heat from the generators is not adequate to serve the thermal load demand, the boiler supplies the rest of the thermal power. For this study, diesel is used as fuel for the boiler, and the boiler has an efficiency of 85%. The HOMER software assumes that the boiler is an existing infrastructure, and its capital cost is not included in the total project expenses. However, the fuel price is added to the project cash-flow.

4. Design Considerations

Several features are considered in this project to estimate an accurate techno-economic evaluation of N-R MHES. Although numerous KPIs can be evaluated in N-R MHES, the three most important KPIs – NPC, COE, and GHG emissions is conducted here as well. It is obvious that the fuel price, load demand, expenses of the equipment, and the efficiency of the machine would not be the same during the entire project life (30 years). Consequently, a multi-year analysis has been performed by varying the most significant parameters that influence the selected KPIs.

4.1. Key Performance Indicators

4.1.1. Net Present Cost

The NPC or life cycle cost of supplies is the present value of all cost (e.g., installation cost, operating cost, replacement cost, fuel expense, emission penalties, purchasing cost of power from the grid), minus the present value of all the revenues; over the project lifetime [34].

The NPC can be calculated from the following equations. Since, the net present value and the net present cost differ only in sign:

$$\text{Net Present Cost (NPC)} = -\text{Net Present Value (NPV)} \tag{8}$$

$$\text{NPV} = \frac{\text{Cash flow}}{(1+i)^t} - \text{Initial Investment} \tag{9}$$

where i and t indicate discount rate (%) and number of time periods, respectively.

If it is intended to calculate the NPC for longer project lifetime with multiple cash flows, the formula can be modified as:

$$\text{NPV} = \sum_{t=0}^{n} \frac{R_t}{(1+i)^t} \tag{10}$$

where R_t and n denote net cash inflow-outflow in unit time period and project lifetime, respectively.

4.1.2. Cost of Energy

The COE is the average cost per unit useful electrical energy, kilowatt-hour (kWh). For a CHP system, the COE is calculated as follows [35]:

$$\text{COE} = \frac{C_{\text{ANNUAL,TOTAL}} - C_{\text{BOILER}} H_{\text{SERVED}}}{H_{\text{SERVED}}} \tag{11}$$

where $C_{\text{ANNUAL, TOTAL}}$, C_{BOILER}, H_{SERVED}, and E_{SERVED} represent total annualized cost of the system ($/year), boiler marginal cost ($/kWh), total annualized thermal load served (kWh/year), annualized total electrical load served (kWh/year), respectively.

The unit of cash can vary from region to region. The second term of the numerator is responsible for serving the thermal load. If the system does not serve the thermal load, then $C_{\text{BOILER}} H_{\text{SERVED}} = 0$.

4.1.3. Greenhouse Gas Emission

The emission of pollutants mostly results from the generation of electric energy from the conventional thermal generator, production of thermal energy from the boiler, and consumption of grid electricity. The pollutants that are usually emitted from the generator, boiler, and electric grid are carbon dioxide, carbon monoxide, particulate matter, sulfur dioxide, and nitrogen oxides.

In the simulation, the emission from the generator and boiler is determined in the same way; whereas, the emission from the grid is calculated differently. The annual emission from generator or boiler is estimated as follows:

$$\begin{aligned}
&\text{Emission of any Pollutant by Generator/Boiler} \\
&= \text{Emission Factor} \left(\tfrac{g}{\text{kWh}} \right) \times \text{Total Annual Fuel Consumption (kWh)}
\end{aligned} \tag{12}$$

Since the grid is capable of purchasing and selling energy, the annual grid emission is estimated by Equation (13):

$$\begin{aligned}
&\text{Emission of any Pollutant by Grid} \\
&= \text{Emission factor} \left(\tfrac{g}{\text{kWh}} \right) \times (\text{Grid Purchase Energy} - \text{Grid Sale Energy})(\text{kWh})
\end{aligned} \tag{13}$$

To calculate the total emission throughout the project lifetime, the annual emission amount is multiplied by the project lifetime.

4.2. Multi-year Analysis

For a multi-year analysis, six factors are examined that substantially affect the N-R MHES performance. The percentage of changes of the selected factors are noted in Table 7. The value of the parameters increases or decreases gradually according to the changes indicated in Table 7. For example, the grid electricity price is raised by 9.6% in each year throughout the total project lifetime.

Table 7. Parameters of the multi-year analysis.

Criteria	Value (Change in %/year)
Grid Electricity Price Increment	9.6 [36]
PV Degradation	1 [37]
Fuel Price Increment (Uranium)	0.0257 [38]
Fuel Price Increment (Diesel)	10 [39]
Electricity Demand Increment	0.3 [40]
Thermal Demand Increment	0.1 [40]

4.3. Control Algorithm

A modified cycle charging control strategy is implemented in this project. In cycle charging strategy, renewable generators first supply electricity to meet the load demand, and the excess energy, if any, is occupied by the battery pack to charge in between SOC_{max} and SOC_{min}. If the renewable generators are incapable of fully supplying the electricity demand, standby battery bank comes to the scenario and supports the load along with the renewable generators. If the combination of renewable generators and batteries cannot fulfill the electricity demand, a back-up diesel generator will be initiated to supply energy [41].

The modified cycle charging strategy is same as "cycle charging" strategy except for the inclusion of the MMR. The MMR provides a continuous supply of electricity and is turned-on always throughout the year, if there is no forced shutdown of MMR. However, if the energy system does not have any renewable generator and the energy system includes conventional FFTG, battery, and electric grid, the electricity demand will be served by the electric grid first, then by the battery, then by the thermal generator.

Figure 6 demonstrates the working strategy of the modified cycle charging, which is applied in this study. Initially, the renewable generators, along with MMR, will be initiated to support the load. If there is any excess energy, it will be applied to charge the battery, and the rest of the electric energy, if any, will be sold to the grid. In this circumstance, S_2, S_3, S_4, S_7, S_8, S_9, and S_{10} will be closed. If the electricity demand is increased slightly, the grid will supply the required electricity to the load. In this state, S_2, S_3, S_4, S_7, S_8, S_9, and S_{10} will also be turned off. The further rise in the demand, the battery bank will be operated; so, S_2, S_3, S_4, S_5, S_7, S_8, S_9, and S_{10} will be closed. If the electricity demand is increased more, the diesel generator (if any) will be started. Therefore, S_1, S_2, S_3, S_4, S_5, S_6, S_7, S_8, S_9, and S_{10} will be operated. It should be noted that S_{10} is a bi-directional switch. For off-grid application, the excess energy will be consumed by dump load. A thermal load controller, rated as 1300 kW, is used in the system to covert the excess electricity into thermal energy to support the thermal load demand.

In a particular case, when the system has no renewable generators, the diesel generator will be initiated to charge the battery, and the grid will start supplying electricity to the load simultaneously (S_1, S_6, S_7, and S_{10} will be closed). If the stand-alone gird is inadequate to accomplish the electricity demand, the battery will be discharged to satisfy the electric load (S_1, S_5, S_6, S_7, and S_{10} will be closed). If the combination of the grid and battery fails to satisfy the electricity demand completely, the diesel generator will be turned on. The diesel generator will always make sure that the battery bank is fully charged. The above-mentioned control algorithm is outlined in Figure 7.

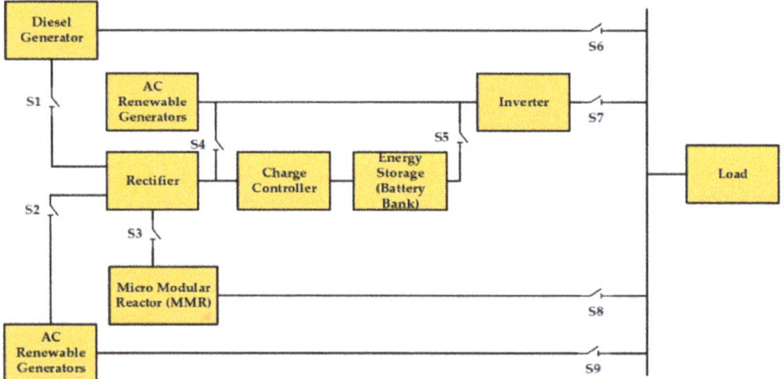

Figure 6. Working strategy of the control algorithm.

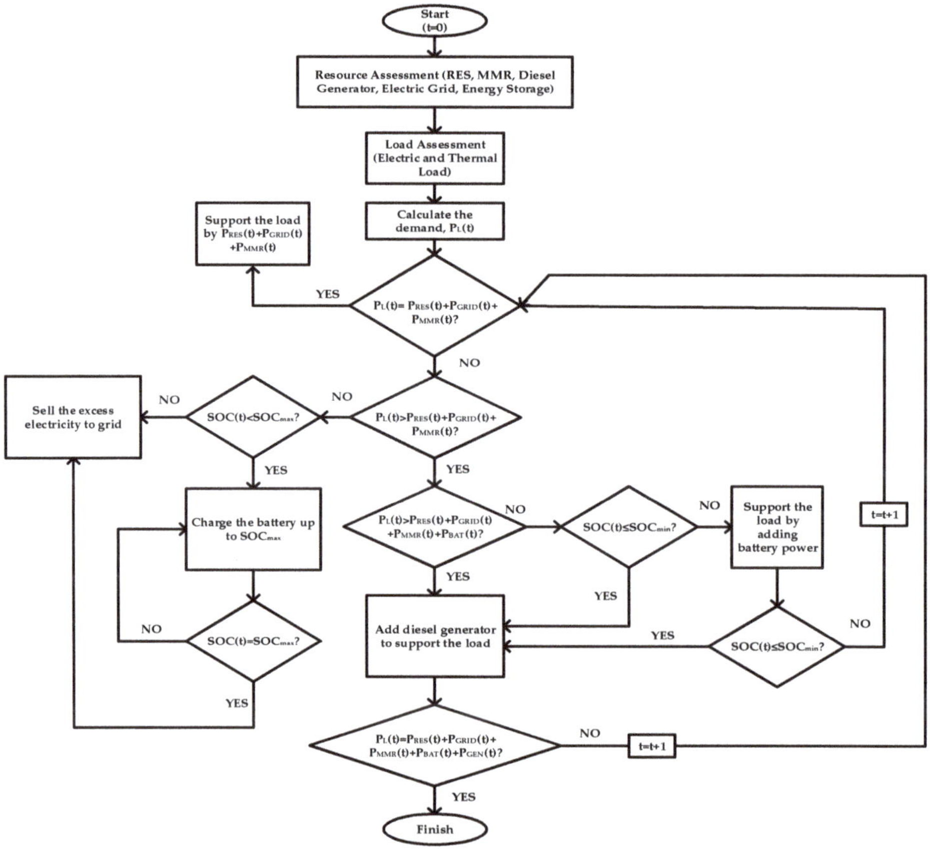

Figure 7. Control algorithm.

5. Simulations

In this section, five different types of energy systems are simulated in the HOMER software. The simulations include several components, such as electric load, thermal load, solar power, wind power, hydro power, energy storage (battery), electric grid, diesel generator, boiler, and power electronics devices — various combinations of these components model the five different energy systems. The equipment ratings are assumed to be the same for all cases.

5.1. Case-01: Conventional Small-scale Fossil Fuel -based Thermal Energy System

In this case, no renewable generator is considered; this system model consists of electric load, thermal load, electric grid, battery, and boiler. The conventional small-scale grid-connected thermal energy system is the most common electrical power generation system arrangement. This scheme is considered for the sake of comparison with N-R MHES. The equipment ratings are previously stated in Section 3. Figure 8 shows the system configuration.

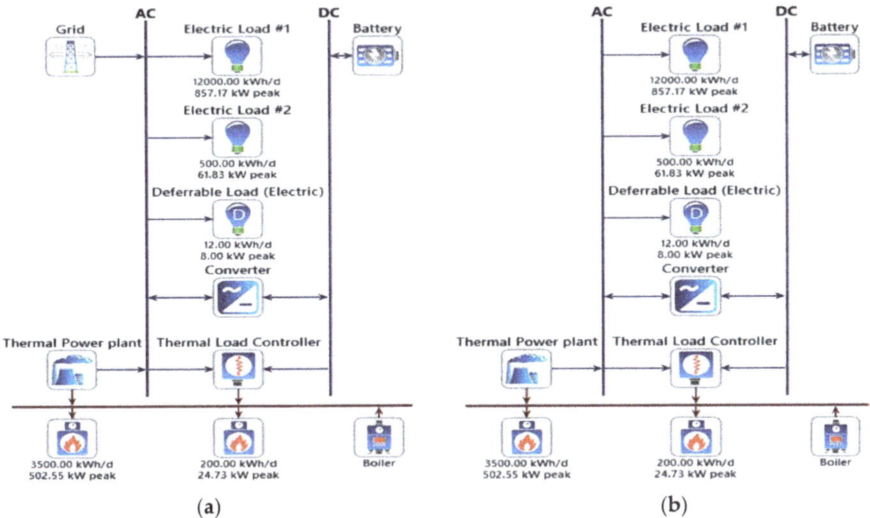

Figure 8. (a) Conventional small-scale grid-connected fossil fuel-based thermal energy system; (b) Conventional small-scale Off-grid fossil fuel-based thermal energy system.

Table 8 presents the total electricity and thermal energy production and consumption scenario in the grid-connected system for the first year of the project period. Since the generated electricity and thermal energy production/distribution pattern are similar for each year of the project lifetime, only the first-year generation/consumption data of the five cases have been investigated in this paper. According to the control algorithm, the electric load demand is mostly fulfilled by the electric grid purchased energy (99.99) and the genset supplies very small amount of energy (0.09%) for a short time. As the electric grid has an adequate power rating, it is capable of supplying electricity sufficiently to the load. It should be remarked that the grid is always competent in handling any size of load demand; but, if the grid exceeds its' rated capacity, the "demand charge" is applied to the user – causing a significantly high electricity bill for the customer. HOMER always optimizes the system to the most economical one; hence, the battery bank is operated to avoid the demand charge in this case.

Since the genset is operated for a short duration, the amount of cogenerated heat is very small. Moreover, there is no other thermal power generation sources in the system except genset and boiler. Therefore, 99.98% thermal demand is accomplished by the boiler. Table 9 presents the energy production and consumption profile of the off-grid system for case-01. The Genest fully meets the electric demand. The thermal demand is accomplished mostly by the cogenerated heat of genset.

Table 8. Energy production and consumption scenario of grid-connected mode for the first year of project life (case-01).

Electricity Production	Amount (%)	Electricity Consumption	Amount (%)	Thermal Energy Production	Amount (%)	Thermal Energy Consumption	Amount (%)
Grid Purchase	99.99	Primary Load	99.91	Boiler	99.98	Thermal load	100
Genset	0.01	Deferrable Load	0.09	Genset	0.02		
Total	100	Total	100	Total	100	Total	100

Table 9. Energy production and consumption scenario of off-grid mode for the first year of project life (case-01).

Electricity Production	Amount (%)	Electricity Consumption	Amount (%)	Thermal Energy Production	Amount (%)	Thermal Energy Consumption	Amount (%)
Genset	100	Primary Load	99.91	Genset	97.42	Thermal load	100
		Deferrable Load	0.09	Boiler	2.57		
				Excess Electricity	0.01		
Total	100	Total	100	Total	100	Total	100

5.2. Case-02: Small-scale Stand-alone Renewable Energy Sources-based Energy System

This is a RES-based energy system with the electric grid; no thermal generator is installed in this case. The boiler is included to serve the thermal load. The RESs include wind, solar, and hydro power in this case. Figure 9 shows the stand-alone grid-connected RES-based energy system.

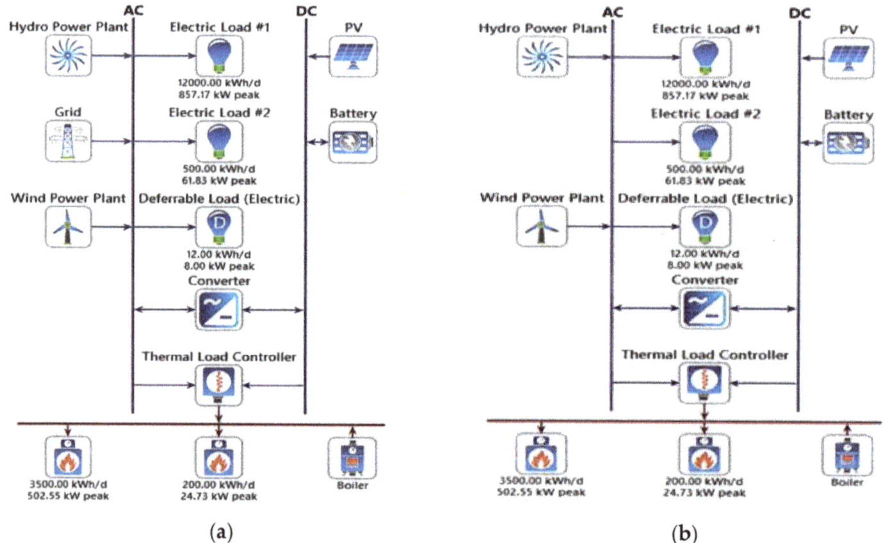

Figure 9. (a) Small scale stand-alone grid-connected renewable energy sources-based energy system; (b) Small scale stand-alone Off-grid renewable energy sources-based energy system.

Since the electricity production by RESs is insignificant compared to the electricity demand and the availability of RESs is irregular, a substantial amount of electricity (45.90%) is purchased from the grid, addressed in Table 10. The rest of the electricity demand is fulfilled by the combination of wind (31.10%), hydro (14.90%), and solar (8.10%). The primary electric load is the summation of Electric Load #1 and Electric Load #2.

The thermal demand is totally met by boiler since boiler is the only thermal power generator in this system architecture. The thermal load is also the summation of Thermal Load #1 and Thermal load #2. In case-02, the system equipment cannot fulfill the electric and thermal load demand without electric grid. Even the battery bank is not capable of fulfilling the electricity demand in case-02 (off-grid mode). Hence, there is no feasible energy distribution scenario for case-02 (off-grid mode).

Table 10. Energy production and consumption scenario of grid-connected mode for the first year of project life (case-02).

Electricity Production	Amount (%)	Electricity Consumption	Amount (%)	Thermal Energy Production	Amount (%)	Thermal Energy Consumption	Amount (%)
Grid Purchase	39.30	Primary Load	94.70				
Wind	27.60	Grid Sales	5.19	Boiler	100	Thermal Load	100
Hydro	22.60	Deferrable Load	0.11				
Solar	10.50						
Total	100	Total	100	Total	100	Total	100

5.3. Case-03: Conventional Small-scale Fossil Fuel-based Thermal and Renewable Energy Sources-based Hybrid Energy System

Figure 10 outlines a typical fossil fuel-based thermal-renewable micro hybrid energy system, which is a combination of case-01 and case-02. The power rating and costs are the same as the previous specification for all components.

Table 11 presents the amount of the electric and thermal energy production and consumption within the system. The contribution of electricity production by the sources is same as case-02. Though the diesel Genset is available in the system, HOMER optimizes the system for the lowest cost and prefers the electric grid over the Diesel genset to serve the electricity demand. Therefore, the electric demand is fulfilled without operating the genset. Moreover, the RESs are non-dispatchable and generates excess energy—higher than the need—for a definite period depending on the availability of RESs; this small amount of surplus electricity (6.19%) is sold to the grid, indicated in Table 11. As the genset is not operated at all, no cogenerated heat is available for this system. Hence, the thermal demand is fully supplied by the boiler.

In the off-grid mode, 46.47% and 75.70% of the total electric and thermal demand, respectively, are satisfied by the genset and cogenerated heat of the genset. Around 15.3% of the thermal demand is met by the surplus electric power, which is generated by wind, hydro, and solar. The thermal load controller, depicted in Figure 10b, converts the excess electric power into thermal power if needed by the thermal load. For this case, energy production and consumption scenario of off-grid mode for the first year of project have shown in the Table 12.

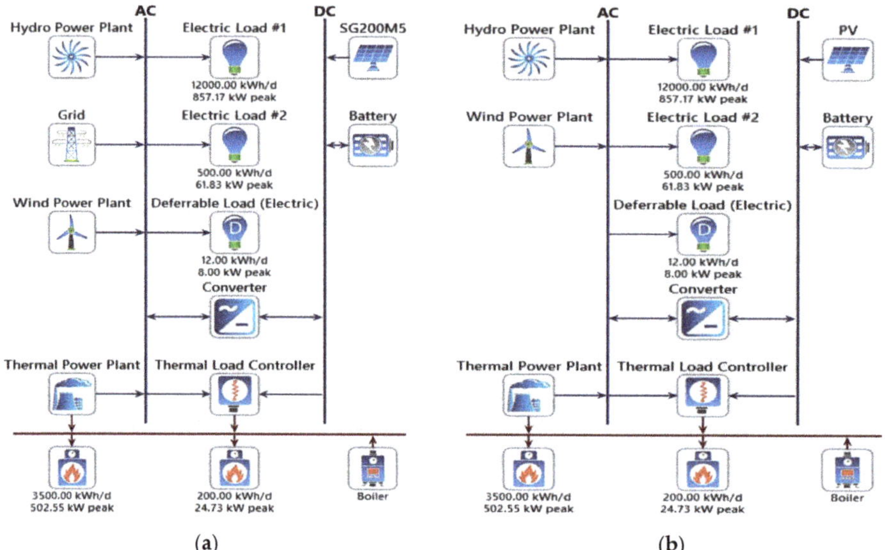

Figure 10. (**a**) Conventional small-scale fossil fuel-based grid-connected thermal and renewable energy sources-based hybrid energy system; (**b**) Conventional small-scale fossil fuel-based off-grid thermal and renewable energy sources-based hybrid energy system.

Table 11. Energy production and consumption scenario of grid-connected mode for the first year of project life (case-03).

Electricity Production	Amount (%)	Electricity Consumption	Amount (%)	Thermal Energy Production	Amount (%)	Thermal Energy Consumption	Amount (%)
Grid Purchase	39.30	Primary Load	94.70				
Wind	27.60	Grid Sales	5.19	Boiler	100	Thermal Load	100
Hydro	22.60	Deferrable Load	0.11				
Solar	10.50						
Total	100	Total	100	Total	100	Total	100

Table 12. Energy production and consumption scenario of off-grid mode for the first year of project life (case-03).

Electricity Production	Amount (%)	Electricity Consumption	Amount (%)	Thermal Energy Production	Amount (%)	Thermal Energy Consumption	Amount (%)
Genset	56.86	Primary Load	99.90	Genset	87.46		
Wind	14.3			Excess Electricity	2.77	Thermal Load	100
Hydro	23.4	Deferrable Load	0.10				
Solar	5.44			Boiler	9.77		
Total	100	Total	100	Total	100	Total	100

5.4. Case-04: Small-scale Stand-alone Nuclear Energy System

Figure 11 illustrates a stand-alone grid-connected small-scale nuclear power system. The system configuration is similar to case-01, except the Diesel genset is replaced by an MMR with the same power and CHP rating.

The MMR always provides a continuous supply of electricity at its' rated capacity, and a single unit of MMR is capable of supplying full electricity demand (100%) in the case-04, indicated in Table 13. Table 13 also shows that a considerable amount of surplus energy (47.90%) is being sold to the grid; this ultimately causes profit for the system. Moreover, since the MMR has the 40% CHP rating and the MMR runs continuously throughout the year, the cogenerated heat from MMR is fully sufficient to serve the thermal load demand of the system, depicted in Table 11. The boiler is not contributing in this case to serve the thermal load.

As shown in Table 14, in off-grid mode, the total electric demand is fulfilled by the MMR. The thermal demand is partially (53.9%) met by the cogenerated heat of MMR, and the excess electricity accomplishes the rest (46.1%) of the demand.

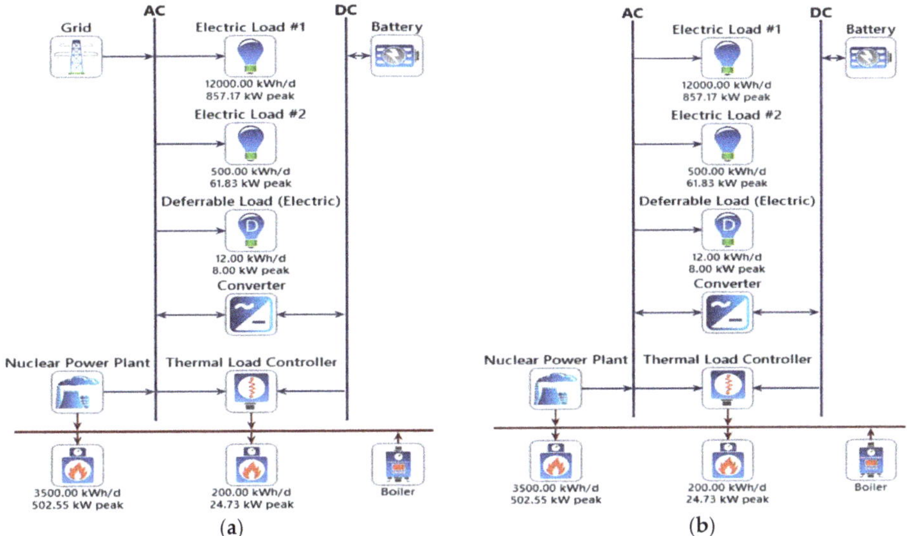

Figure 11. (a) Small-scale grid-connected stand-alone nuclear energy system; (b) Small-scale off-grid stand-alone nuclear energy system.

Table 13. Energy production and consumption scenario of grid-connected mode for the first year of project life (case-04).

Electricity Production	Amount (%)	Electricity Consumption	Amount (%)	Thermal Energy Production	Amount (%)	Thermal Energy Consumption	Amount (%)
		Primary Load	52.05				
MMR	100	Grid Sales	47.90	MMR	100	Thermal Load	100
		Deferrable Load	0.05				
Total	100	Total	100	Total	100	Total	100

Table 14. Energy production and consumption scenario of off-grid mode for the first year of project life (case-04).

Electricity Production	Amount (%)	Electricity Consumption	Amount (%)	Thermal Energy Production	Amount (%)	Thermal Energy Consumption	Amount (%)
MMR	100	Primary Load	99.10	MMR	53.9	Thermal Load	100
		Deferrable Load	0.01	Excess Electricity	46.1		
Total	100	Total	100	Total	100	Total	100

5.5. Case-05: Nuclear-Renewable Micro Hybrid Energy System

This case represents an N-R MHES, similar to case-03, except the diesel generator is interchanged by an MMR. Figure 12 shows the schematic of an N-R MHES.

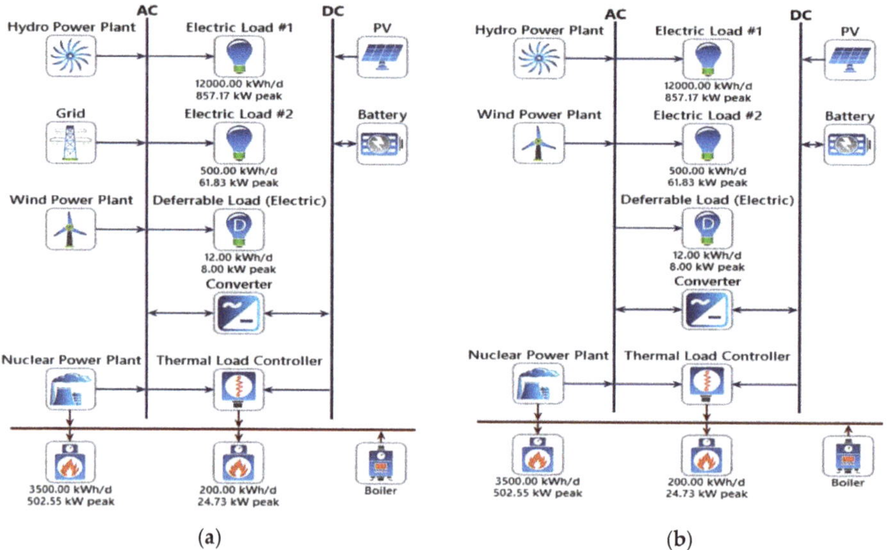

Figure 12. (a) Grid-connected nuclear-renewable micro hybrid energy system; (b) Off-grid nuclear-renewable micro hybrid energy system.

In N-R MHES, three-quarters (76.8%) of the total electricity demand is served by the MMR, and the rest of the electricity demand is achieved by the RESs, mentioned in Table 15. Since electricity production also is massive in this case, a substantial amount of excess energy (59.86%) is sold to the grid. Likewise, the case-04, the whole thermal demand is met by the cogenerated heat from MMR in this case; the boiler is kept in idle mode.

In off-grid mode, 76.8% electric load is served by the MMR which has been shown in Table 16. A significant amount of excess electric energy (58.2% of total thermal demand) generated by the combination of MMR and RESs is used to serve the thermal by the help of thermal load controller.

Table 15. Energy production and consumption scenario of grid-connected mode for the first year of project life (case-05).

Electricity Production	Amount (%)	Electricity Consumption	Amount (%)	Thermal Energy Production	Amount (%)	Thermal Energy Consumption	Amount (%)
MMR	74.90	Primary Load	39.10				
Wind	11.40	Grid Sales	60.86	MMR	100	Thermal Load	100
Hydro	9.37	Deferrable Load	0.04				
Solar	4.33						
Total	100	Total	100	Total	100	Total	100

Table 16. Energy production and consumption scenario of off-grid mode for the first year of project life (case-05).

Electricity Production	Amount (%)	Electricity Consumption	Amount (%)	Thermal Energy Production	Amount (%)	Thermal Energy Consumption	Amount (%)
MMR	81.25	Primary Load	99.90	Excess Electricity	55.9		
Wind	6.19					Thermal Load	100
Hydro	10.2	Deferrable Load	0.10	MMR	44.1		
Solar	2.36						
Total	100	Total	100	Total	100	Total	100

6. Results

The study has investigated three important KPIs of N-R MHES. To evaluate the KPIs, the N-R MHES is compared with four other different energy systems (from case-01 to case-04). The computed NPC, for both grid-connected and off-grid mode, of the five cases are recorded in Table 17. The HOMER Pro software optimizes the system based on the NPC of the system.

For the grid-connected mode, case-01 has the highest NPC and COE, whereas case-04 and case-05 has the lowest NPC and COE, respectively. In case-01 (grid-connected mode), 99.99% electricity of the total electric demand is purchased from the grid. Furthermore, a massive amount of Diesel fuel (161,438 L/year) is utilized in the boiler to serve the thermal load that raises the total system cost. Consequently, the NPC and COE are quite high for this case. Case-04 (grid-connected mode) has the lowest NPC due to lower capital cost and operating cost. In case-05 (grid-connected mode), a large number of surplus electric energy is being generated by the combination of MMR and RES, and the system makes a significant amount of revenue by selling the excess energy to the grid. Besides, the cogenerated heat from MMR is adequate to serve the thermal demand fully; thus, no additional fuel is required to serve the thermal load. Therefore, COE is the lowest for this case. The NPC of case-05 (grid-connected mode) is slightly higher than case-04 (grid-connected mode) due higher initial cost of different RES equipment, such as wind turbine, hydro turbine, and PV panel.

In the off-grid mode of operation, case-01 exhibits the highest NPC and COE. On the other hand, case-04 (off-grid mode) has the lowest NPC and COE. It should be remarked that case-02 does not provide any feasible solution, implying that the system cannot support the electric demand studied in this paper. In case-01 (off-grid mode), the genset supplies the total electric load by burning fuel, which increases the system cost significantly. However, boiler fuel cost is less in this case since the boiler provides only 2.57% of thermal demand. In case-04, the NPC is lowest because the system consists of only MMR; no RESs are involved here. The absence of RESs reduces the system cost significantly as there is no high initial cost.

Furthermore, the substantial excess electric energy is utilized by the thermal load controller to serve the thermal load. Since the boiler is not operated, and there is no additional fuel cost to fulfill the thermal demand, the NPC and COE of the overall system are reduced in case-04 (off-grid mode). Nevertheless, it should be mentioned that a vast amount of excess electric energy (6,210,660 kWh/year) and thermal energy (9,765,760 kWh/year) is generated in case-05 (off-grid mode) compared to the case-04 (off-grid mode). The excess electric and thermal energy produced in case-04 (off-grid mode) are estimated as 4,193,123 kWh/year and 7,748,223 kWh/year, respectively. Due to the absence of electric grid in case-05 (off-grid mode), the excess electric energy is not being sold to the grid; hence, both NPC and COE are higher in case-05 (off-grid mode) than case-04 (off-grid mode).

Table 17. Case-wise comparison of NPC and COE.

Scenario	System Status	NPC (USD)	COE (USD/kWh)
Case-01	Grid-connected	31,583,960	0.3781
	Off-grid	57,040,910	0.7656
Case-02	Grid-connected	19,877,210	0.1907
	Off-grid	NFS *	NFS *
Case-03	Grid-connected	20,142,290	0.1946
	Off-grid	37,807,960	0.4728
Case-04	Grid-connected	10,960,580	0.0345
	Off-grid	12,371,000	0.0856
Case-05	Grid-connected	10,987,980	0.0262
	Off-grid	13,221,380	0.0985

* NFS = Non-feasible Solution.

Table 18 presents the amount of GHG emissions in various cases. In this project, five types of GHGs are studied: carbon dioxide, carbon monoxide, sulfur dioxide, nitrogen oxide, and particulate matter. The particulate matter includes smoke, soot, and liquid droplets. For grid-connected mode, in case-01, all types of GHG emissions are highest because 99.99% of the electric demand is supplied by the electric grid in case-01 and the burning of natural gas produces the grid electricity. As there is no burning of fossil fuel in case-04 (both grid-connected and off-grid mode) and case-05 (both grid-connected and off-grid mode), and the MMR fulfills the demand (both electric and thermal) implicitly, GHG emissions are zero for these cases. It should be mentioned that the amount of GHG emissions is only accounted for the electricity production process in this study; the carbon footprint for construction of the site and equipment manufacturing (i.e., PV panel) are not reflected here. Surprisingly, the GHG emissions are significant in case-02 (grid-connected mode) and case-03 (grid-connected mode). Generally, the GHG emission should be minimal in both cases because these are RES-based grid-connected hybrid energy system, and only the fossil fuel is burnt to support the thermal demand. However, since the amount of purchased electricity from the grid is extensive in these cases and the grid electricity is being generated by burning natural gas, case-02 (grid-connected mode) and case-03 (grid-connected mode) cause a significant amount of GHG emissions.

For off-grid mode, case-01 also produces the maximum emissions due to the burning of diesel in the Genset and the boiler. Case-04 and case-05 show zero-emission since the full demand (both electrical and thermal) is satisfied by the combination of MMR, RESs, and cogenerated heat from MMR.

Table 18. Case-wise comparison of GHG emission.

Pollutant	System Status	Case-01	Case-02	Case-03	Case-04	Case-05
Carbon Dioxide (kg/year)	Grid-connected	1,254,395	771,827	771,827	0	0
	Off-grid	3,261,052	NFS	1,700,908	0	0
Carbon Monoxide(kg/year)	Grid-connected	284	118	118	0	0
	Off-grid	16,742	NFS	8468	0	0
Sulfur Dioxide (kg/year)	Grid-connected	1067	1062	1062	0	0
	Off-grid	7972	NFS	4160	0	0
Nitrogen Oxide (kg/year)	Grid-connected	651	271	271	0	0
	Off-grid	3209	NFS	1623	0	0
Particulate Matter (kg/year)	Grid-connected	49.3	20.6	20.6	0	0
	Off-grid	143	NFS	72.4	0	0

However, the discussion mentioned above cannot draw a single conclusion explicitly since this techno-economic study depends on several variables. Therefore, a sensitivity analysis is also carried out to identify the impact of three critical parameters- discount rate, inflation rate, and project lifetime – on the system performance. Three different values are taken for each parameter, and the system is evaluated based on these values, presented in Table 19. From Table 19, the stand-alone off-grid RES-based system (case-02) is not feasible despite the higher or lower discount rate, inflation rate, and project lifetime in this study. Case-01 (off-grid mode) and case-04 (off-grid mode) cannot also manage the selected electric and thermal load demand, studied here, for longer project lifetime (60 and 100 years), depicted in Table 19. Moreover, case-03 (off-grid mode) is not capable of fulfilling the studied electric and thermal load for a long project lifetime (100 years). All the system infeasibility happens due to capacity shortage. Since the electric and thermal demand is increasing at a certain percentage each year without increasing the capacities of the resources, the current resources cannot serve the demand fully. The higher inflation rate has a notable impact also in the infeasible system. However, the N-R MHES (case-05, both grid-connected and off-grid mode) shows that it can support the demand always with considerably lower NPC, regardless of the higher or lower discount rate, inflation rate, and project lifetime. Higher inflation rate increases the NPC of case-01, case-02, and case-03 significantly.

For the lower discount rate (3% and 8%) and short project lifetime (30 years), case-04 (off-grid mode) could have better performance than case-05 (off-grid mode) in terms of NPC, for instance, scenario No. 1,4,7,10,13, and 16 in Table 19. Sometimes, the NPC could also be lesser in case-04 (off-grid mode) in the high discount rate (10%) and small project lifetime (30 years), such as scenario No. 19,22, and 25.

For the lower value of project lifetime (30 and 60 years) or smaller value of inflation rate (2% and 4%), case-04 (grid-connected) may also show a slightly lower amount, compared to case-05 (grid-connected), of NPC for a few cases (e.g., scenario No. 10 and 19). However, for the higher value of project lifetime (60 and 100 years), case-05 always provides the most economical NPC.

Table 19. Sensitivity analysis.

No.	Discount Rate	Inflation Rate	Project Lifetime	System Status	NPC				
					Case-01	Case-02	Case-03	Case-04	Case-05
1	3	2	30	Grid-connected	79.4 M *	48.8 M	49.0 M	9.75 M	9.13 M
				Off-grid	144 M	NFS	94.5 M	12.3 M	13.4 M
2	3	2	60	Grid-connected	1.11 B **	687 M	687 M	12.4 M	9.07 M
				Off-grid	NFS ***	NFS	1.42 B	NFS	16.3 M
3	3	2	100	Grid-connected	36.1 B	NFS	22.2 B	729 M	23.0 M
				Off-grid	NFS	NFS	NFS	NFS	19.2 M
4	3	4	30	Grid-connected	121 M	73.8 M	73.8 M	8.61 M	7.49 M
				Off-grid	220 M	NFS	144 M	12.1 M	13.3 M
5	3	4	60	Grid-connected	2.94 B	1.82 B	1.82 B	15.2 M	6.27 M
				Off-grid	NFS	NFS	3.77 B	NFS	19.3 M
6	3	4	100	Grid-connected	208 B	NFS	128 B	4.37 B	96.7 M
				Off-grid	NFS	NFS	NFS	NFS	29.7 M
7	3	5	30	Grid-connected	149 M	91.2 M	91.1 M	7.79 M	6.32 M
				Off-grid	273 M	NFS	178 M	11.9 M	13.1 M
8	3	5	60	Grid-connected	4.81 B	2.99 B	2.99 B	18.4 M	3.70 M
				Off-grid	NFS	NFS	6.19 B	NFS	21.8 M
9	3	5	100	Grid-connected	499 B	NFS	306 B	10.7 B	233 M
				Off-grid	NFS	NFS	NFS	NFS	41.2 M
10	8	2	30	Grid-connected	31.7 M	19.9 M	20.1 M	11.0 M	11.0 M
				Off-grid	57.0 M	NFS	37.8 M	12.4 M	13.2 M
11	8	2	60	Grid-connected	128 M	79.6 M	79.9 M	11.3 M	11.2 M
				Off-grid	NFS	NFS	161 M	NFS	13.8 M
12	8	2	100	Grid-connected	666 M	NFS	411 M	21.0 M	11.4 M
				Off-grid	NFS	NFS	NFS	NFS	13.9 M
13	8	4	30	Grid-connected	45.3 M	28.1 M	28.4 M	10.6 M	10.5 M
				Off-grid	81.9 M	NFS	54.0 M	12.4 M	13.3 M

Table 19. *Cont.*

No.	Discount Rate	Inflation Rate	Project Lifetime	System Status	NPC				
					Case-01	Case-02	Case-03	Case-04	Case-05
14	8	4	60	Grid-connected	296 M	183 M	183 M	11.4 M	10.7 M
				Off-grid	NFS	NFS	374 M	NFS	14.4 M
15	8	4	100	Grid-connected	3.19 B	NFS	1.96 B	66.7 M	11.7 M
				Off-grid	NFS	NFS	NFS	NFS	14.9 M
16	8	5	30	Grid-connected	54.5 M	33.7 M	33.9 M	10.4 M	10.1 M
				Off-grid	98.9 M	NFS	65.0 M	12.4 M	13.4 M
17	8	5	60	Grid-connected	459 M	284 M	284 M	11.6 M	10.3 M
				Off-grid	NFS	NFS	582 M	NFS	14.8 M
18	8	5	100	Grid-connected	7.19 B	NFS	4.43 B	143 M	12.8 M
				Off-grid	NFS	NFS	NFS	NFS	15.7 M
19	10	2	30	Grid-connected	23.2 M	14.8 M	15.0 M	11.2 M	11.3 M
				Off-grid	41.6 M	NFS	27.8 M	12.3 M	13.1 M
20	10	2	60	Grid-connected	63.5 M	39.6 M	39.9 M	11.3 M	11.4 M
				Off-grid	NFS	NFS	78.8 M	NFS	13.4 M
21	10	2	100	Grid-connected	178 M	NFS	110 M	13.3 M	11.5 M
				Off-grid	NFS	NFS	NFS	NFS	13.5 M
22	10	4	30	Grid-connected	32.3 M	20.2 M	20.5 M	10.9 M	11.0 M
				Off-grid	58.1 M	NFS	38.5 M	12.4 M	13.2 M
23	10	4	60	Grid-connected	134 M	83.2 M	83.5 M	11.3 M	11.1 M
				Off-grid	NFS	NFS	168 M	NFS	13.8 M
24	10	4	100	Grid-connected	724 M	NFS	446 M	22.0 M	11.4 M
				Off-grid	NFS	NFS	NFS	NFS	13.9 M
25	10	5	30	Grid-connected	38.4 M	23.9 M	24.2 M	10.8 M	10.7 M
				Off-grid	69.3 M	NFS	45.8 M	12.4 M	13.3 M
26	10	5	60	Grid-connected	201 M	124 M	125 M	11.3 M	10.9 M
				Off-grid	NFS	NFS	253 M	NFS	14.1 M
27	10	5	100	Grid-connected	1.55 B	NFS	953 M	36.4 M	11.4 M
				Off-grid	NFS	NFS	NFS	NFS	14.3 M

* M = million, ** B = billion, *** NFS = No Feasible Solution.

7. Discussion

Large-scale N-R HES is not a new concept. Research and innovation for the expansion of N-R HES are taking place in several countries. Nevertheless, due to significant risk and high capital cost of large-scale NPP, small-scale NPP integration with RESs has been proposed and evaluated in this paper. Although the N-R MHES depends on the availability of the local RESs, the N-R MHES always provides a baseload supply to strengthen the resiliency and stability of the hybrid energy system. This paper is intended to provide a clear idea of the technical and economic aspects of small-scale N-R HES. The key findings of this study can be summarized as follows.

From the sensitivity analysis, it can be concluded that the N-R MHES could be a resilient energy supply source for sustainable energy solutions in the future. The N-R MHES shows a significant benefit for the long-term planning of a reliable energy system. Furthermore, the MMR can be a suitable replacement for a Diesel genset in terms of NPC, COE, and emissions. From case-02 (off-grid mode), it is observed that the off-grid RES-based energy system is rarely capable or not capable of handling large/medium-scale electricity demand adequately. However, the off-grid RES-based energy system can be an impactful solution for small-scale electric/thermal demand. The case-02 (grid-connected mode) shows that the grid-energy is not always a clear form of an electric power source if the grid-energy comes from the burning of fossil fuel, although the grid-energy is deemed to be clean. Grid-connected N-R MHES could be the most cost-effective solution in terms of COE to provide medium/large-scale electric power supply, depicted in case-05 (grid-connected mode). Moreover, as an extensive amount of excess electric and thermal energy is available in case-05 (off-grid mode), off-grid N-R MHES could be a suitable option for medium/large-scale remote industrial applications, such as desalination plants, mining stations, and EV charging platforms. Also, N-R MHES can provide the best solution for decarbonization pathways project since there are no emissions in case-05.

Few other parameters, e.g., availability of RES, extreme conditions of RES, and abrupt change in electric and thermal load demand, also may influence the whole system performance and system stability. But these kinds of severe conditions are not regarded in this study. Furthermore, the carbon tax incorporation with the NPC and COE is beyond this study. If the carbon tax is included in the study, the NPC and COE of the case-01, case-02, and case-03 would be increased further. In addition, grid stability consideration is beyond of the scope of this study.

Author Contributions: Conceptualization, H.A.G., M.R.A. and M.I.A.; methodology, M.R.A. and M.I.A.; software, M.R.A.; validation, H.A.G., M.R.A. and M.I.A.; formal analysis, M.R.A.; investigation, M.R.A.; resources, M.R.A.; data curation, M.R.A.; writing—original draft preparation, M.R.A. and M.I.A.; writing—review and editing, M.R.A. and M.I.A; visualization, H.A.G., M.R.A. and M.I.A.; supervision, H.A.G.; project administration, H.A.G.; funding acquisition, H.A.G. All authors have read and agreed to the published version of the manuscript.

Funding: This research was funded by Natural Sciences and Engineering Research Council of Canada, grant number 210320.

Conflicts of Interest: The authors declare no conflict of interest.

Nomenclature

CCS	Carbon Capture and Storage
CHP	Combined Heat and Power
COE	Cost of Energy
ESS	Energy Storage System
FFTG	Fossil Fuel-based Thermal Generator
GHG	Greenhouse Gas
HES	Hybrid Energy System
HOMER	Hybrid Optimization Model for Electric Renewable
IAEA	International Atomic Energy Agency
IRR	Internal Rate of Return
KPI	Key Performance Indicator

LCOE	Levelized Cost of Energy
SDG	Sustainable Development Goal
STC	Standard Test Condition
MM	Mobile Microgrid
MMR	Micro Modular Reactor
NASA	National Aeronautics and Space Administration
NPC	Net Present Cost
NPP	Nuclear Power Plant
NPV	Net Present Value
N-R HES	Nuclear-Renewable Hybrid Energy System
N-R MHES	Nuclear-Renewable Micro Hybrid Energy System
PV	Photovoltaic
RES	Renewable Energy Source
SMR	Small Modular Reactor
SOC_{max}	State of Charge (maximum)
SOC_{min}	State of charge (minimum)
TCI	Total Capital Investment
vSMR	Very small reactors

References

1. Goal 7: Affordable & Clean Energy—UN SDG. Available online: https://www.gvicanada.ca/goal-7-affordable-and-clean-energy/ (accessed on 10 December 2019).
2. Usman, M.; Khan, M.T.; Rana, A.S.; Ali, S. Techno-economic analysis of hybrid solar-diesel-grid connected power generation system. *J. Electr. Syst. Inf. Technol.* **2018**, *5*, 653–662. [CrossRef]
3. Freris, L.; Infield, D. *Renewable Energy in Power Systems*; John Wiley & Sons, Ltd.: Hoboken, NJ, USA, 2008.
4. Conserve Energy Future What Is Sustainable Energy and Its Types. Available online: https://www.conserve-energy-future.com/sustainableenergy.php (accessed on 9 December 2019).
5. Bragg-Sitton, S.M.; Boardman, R.; Rabiti, C.; Suk Kim, J.; McKellar, M.; Sabharwall, P.; Chen, J.; Cetiner, M.S.; Harrison, T.J.; Qualls, A.L. *Nuclear-Renewable Hybrid Energy Systems: 2016 Technology Development Program Plan*; Oak Ridge National Lab.: Oak Ridge, TN, USA, 2016; p. 1333006.
6. Ruth, M.F.; Zinaman, O.R.; Antkowiak, M.; Boardman, R.D.; Cherry, R.S.; Bazilian, M.D. Nuclear-renewable hybrid energy systems: Opportunities, interconnections, and needs. *Energy Convers. Manag.* **2014**, *78*, 684–694. [CrossRef]
7. IAEA TECDOC No: 1885. Nuclear-renewable hybrid energy systems for decarbonized energy production and cogeneration. In Proceedings of the Technical Meeting, Vienna, Austria, 22–25 October 2018; Technical Meeting on Nuclear-Renewable Hybdrid Energy Systems for Decarbonized Energy Production and Cogeneration, International Atomic Energy Agency. IAEA: Vienna, Austria, 2019. ISBN 978-92-0-161419-3.
8. Boldon, L.; Sabharwall, P.; Bragg-Sitton, S.; Abreu, N.; Liu, L. Nuclear Renewable Energy Integration: An Economic Case Study. *Int. J. Energy Environ. Econ* **2015**, *28*, 85–95.
9. Suman, S. Hybrid nuclear-renewable energy systems: A review. *J. Clean. Prod.* **2018**, *181*, 166–177. [CrossRef]
10. Ruth, M.; Cutler, D.; Flores-Espino, F.; Stark, G.; Jenkin, T. *The Economic Potential of Three Nuclear-Renewable Hybrid Energy Systems Providing Thermal Energy to Industry*; National Renewable Energy Lab.: Golden, CO, USA, 2016; p. 1335586.
11. Giatrakos, G.P.; Tsoutsos, T.D.; Mouchtaropoulos, P.G.; Naxakis, G.D.; Stavrakakis, G. Sustainable energy planning based on a stand-alone hybrid renewableenergy/hydrogen power system: Application in Karpathos island, Greece. *Renew. Energy* **2009**, *34*, 2562–2570. [CrossRef]
12. U-Battery What Is U-Battery? Available online: https://www.u-battery.com/what-is-u-battery (accessed on 9 December 2019).
13. Westinghouse Nuclear. New Plants: eVinci™ Micro Reactor. Available online: http://www.westinghousenuclear.com/new-plants/evinci-micro-reactor (accessed on 9 December 2019).
14. Gabbar, H.A.; Abdussami, M.R. Feasibility Analysis of Grid-Connected Nuclear-Renewable Micro Hybrid Energy System. In Proceedings of the 2019 IEEE 7th International Conference on Smart Energy Grid Engineering (SEGE), Oshawa, ON, Canada, 12–14 August 2019; pp. 294–298.

15. Mishra, M.; Saxena, N.K.; Mishra, P. ANN Based AGC for Hybrid Nuclear-Wind Power System. In Proceedings of the 2016 International Conference on Micro-Electronics and Telecommunication Engineering (ICMETE), Ghaziabad, India, 22–23 September 2016; pp. 410–415.
16. Elma, O.; Gabbar, H.A. Design and Analysis of Mobile Hybrid Energy System for Off-Grid Applications. In Proceedings of the 2019 International Conference on Power Generation Systems and Renewable Energy Technologies (PGSRET), Istanbul, Turkey, 26–27 August 2019; pp. 1–6.
17. Abdussami, M.R.; Gabbar, H.A. Flywheel-based Micro Energy Grid for Reliable Emergency Back-up Power for Nuclear Power Plant. In Proceedings of the 2019 International Conference on Smart Energy Systems and Technologies (SEST), Porto, Portugal, 9–11 September 2019; pp. 1–6.
18. Finding Data to Run HOMER. Available online: https://www.homerenergy.com/products/pro/docs/latest/finding_data_to_run_homer.html (accessed on 26 December 2019).
19. Real Discount Rate. Available online: https://www.homerenergy.com/products/pro/docs/latest/real_discount_rate.html (accessed on 5 January 2020).
20. GlobalPetrolPrices.com Canada Diesel Prices. 18 November 2019. Available online: https://www.globalpetrolprices.com/Canada/diesel_prices/ (accessed on 25 November 2019).
21. How HOMER Calculates the PV Array Power Output. Available online: https://www.homerenergy.com/products/pro/docs/latest/how_homer_calculates_the_pv_array_power_output.html (accessed on 25 November 2019).
22. How HOMER Calculates Wind Turbine Power Output. Available online: https://www.homerenergy.com/products/pro/docs/latest/how_homer_calculates_wind_turbine_power_output.html (accessed on 25 November 2019).
23. How HOMER Calculates the Hydro Power Output. Available online: https://www.homerenergy.com/products/pro/docs/latest/how_homer_calculates_the_hydro_power_output.html (accessed on 25 November 2019).
24. Morales Pedraza, J. *Small Modular Reactors for Electricity Generation*; Springer: Cham, Switzerland, 2017; ISBN 978-3-319-52215-9.
25. World Nuclear Association. Small Nuclear Power Reactors. Available online: https://www.world-nuclear.org/information-library/nuclear-fuel-cycle/nuclear-power-reactors/small-nuclear-power-reactors.aspx (accessed on 8 December 2019).
26. U-Battery. Available online: https://www.u-battery.com/ (accessed on 31 December 2019).
27. Allen, K.S.; Hartford, S.K.; Merkel, G.J. Feasibility Study of a Micro Modular Reactor for Military Ground Applications. *J. Def. Manag.* **2018**, *8*. [CrossRef]
28. USNC. Available online: https://usnc.com/ (accessed on 9 December 2019).
29. T. Abram *et al.*, "Design of a U-Battery®," p. 23, Nov. 2011. Available online: https://www.u-battery.com/media (accessed on 31 March 2020).
30. Abram, T.; Marsden, B.; Wickham, T.; Ding, M.; Kloosterman, J.L.; Kooijman, T.; Linssen, R. Design of a U-Battery®. *Delft Tech. Univ.* **2011**, 23.
31. Nuclear Power Economics|Nuclear Energy Costs—World Nuclear Association. Available online: https://www.world-nuclear.org/information-library/economic-aspects/economics-of-nuclear-power.aspx (accessed on 20 March 2020).
32. Natural Gas 1998, Issues and Trends. Available online: https://www.eia.gov/naturalgas/archive/056098.pdf (accessed on 27 December 2019).
33. Ontario Time-of-Use Electricity Rates. Available online: http://www.ontario-hydro.com/current-rates (accessed on 27 December 2019).
34. Arnaboldi, M.; Azzone, G.; Giorgino, M. Long-and Short-Term Decision Making. In *Performance Measurement and Management for Engineers*; Elsevier: Amsterdam, The Netherlands, 2015; pp. 107–115. ISBN 978-0-12-801902-3.
35. Levelized Cost of Energy. Available online: https://www.homerenergy.com/products/pro/docs/latest/levelized_cost_of_energy.html (accessed on 27 December 2019).
36. Jackson, T.; Stedman, A.; Aliakbari, E.P.; Green, K. *Evaluating Electricity Price Growth in Ontario*; FRASER Institute: Vancouver, BC, Canada, 2017.
37. Commercial Solar Panel Degradation: What You Should Know and Keep in Mind. Available online: https://businessfeed.sunpower.com/articles/what-to-know-about-commercial-solar-panel-degradation (accessed on 25 November 2019).

38. *U.S Department of Energy; 2018 Uranium Marketing Annual Report*; U.S. Energy Information Administration: Washington, DC, USA, 2018; Volume 70.
39. Erwin, S. New marine fuel rules to boost diesel prices for at least a year: Analysts. *Reuters* **2019**. Available online: https://www.reuters.com/article/us-imo-shipping-usa-refiners/new-marine-fuel-rules-to-boost-diesel-prices-for-at-least-a-year-analysts-idUSKCN1TB2CJ (accessed on 31 March 2020).
40. Canada's Energy Future 2018, An Energy Market Assessment. National Energy Board. Available online: https://www.cer-rec.gc.ca/nrg/ntgrtd/ftr/2018/2018nrgftr-eng.pdf (accessed on 31 March 2020).
41. Upadhyay, S.; Sharma, M.P. Development of hybrid energy system with cycle charging strategy using particle swarm optimization for a remote area in India. *Renew. Energy* **2015**, *77*, 586–598. [CrossRef]

Article

Fault Current Tracing and Identification via Machine Learning Considering Distributed Energy Resources in Distribution Networks [†]

Wanghao Fei and Paul Moses *

School of Electrical and Computer Engineering, University of Oklahoma, Norman, OK 73019, USA; whf@ou.edu
* Correspondence: pmoses@ou.edu
† This paper is an extended version of our paper published in In Proceedings of the 2019 IEEE 7th International Conference on Smart Energy Grid Engineering (SEGE), Oshawa, ON, Canada, 12–14 August 2018; pp. 196–200.

Received: 11 October 2019; Accepted: 11 November 2019; Published: 14 November 2019

Abstract: The growth of intermittent distributed energy sources (DERs) in distribution grids is raising many new operational challenges for utilities. One major problem is the back feed power flows from DERs that complicate state estimation for practical problems, such as detection of lower level fault currents, that cause the poor accuracy of fault current identification for power system protection. Existing artificial intelligence (AI)-based methods, such as support vector machine (SVM), are unable to detect lower level faults especially from inverter-based DERs that offer limited fault currents. To solve this problem, a current tracing method (CTM) has been proposed to model the single distribution feeder as several independent parallel connected virtual lines that traces the detailed contribution of different current sources to the power line current. Moreover, for the first time, the enhanced current information is used as the expanded feature space of SVM to significantly improve fault current detection on the power line. The proposed method is shown to be sensitive to very low level fault currents which is validated through simulations.

Keywords: current tracing; fault current; distributed energy resources; network model

1. Introduction

Due to the increasing penetration rate of distributed energy sources (DERs), such as solar power injected at the distribution side of the power system, during the past decade, distribution grids have become large, complex, interconnected networks. The infusion of DERs onto distribution grids raises some new state estimation challenges, such as fault current identification, which was less of an issue in conventional distribution grids operating without DERs.

The traditional distribution grid was not affected by irregular back power feeds caused by intermittent DER activity, as it was originally designed for single direction power flow from the substation to the customers. The ordinary fault current identification process is based on the fault current threshold where no DERs are injected into the distribution grid. With DERs, excess power can be generated and consumed by the customer or feed back to the distribution feeder. This raises new challenges, especially in grid modeling methods for fault current identification, where the impact of back feed power flow has to be considered for determining the fault current threshold. Failure to do so may have far reaching and costly consequences such as inadvertent tripping of circuits or overlooking faults in the system.

Researchers have proposed many grid modeling methods for different types of power grids such as averaged models [1,2] and the port-Hamiltonian-based dynamic power system model [3]. Certain modeling methods are suited for specific parts of the grid such as cyber-physical power systems framework modeling [4,5], topology modeling [6,7], load flow modeling [8,9], lightning protection models [10], and traveling wave modeling [11]. In [12,13], the author used pattern recognition to identify the fault current with the waveform data, which is an efficient tool for power transmission line fault detection. In addition, with the benefit of fast developing artificial intelligence (AI) technologies, many AI-based power system fault identification methods have been proposed as well, such as support vector machine (SVM) [14], k-nearest neighbors algorithm (KNN) [13], pattern recognition [15], and deep learning [16].

Although different AI methods have been used to improve the accuracy of power system fault identification, the approaches still heavily rely on grid modeling methods for some part of the power grid. Most importantly, almost all of these AI-based fault current identification methods based on existing grid modeling methods totally depend on the existing power line infrastructure, in that currents from different power sources are congested to one end of a single power line and flow towards the other end [17–19]. These methods are very effective when it comes to the identification of larger fault currents, but some detailed current information such as those from inverter-based DERs are difficult to detect. A much more detailed grid modeling method is needed to augment the AI method to be more sensitive to detect some lower level faults. This is particularly important in modern distribution grids, as inverter-based DERs are known to produce exceptionally small fault current contributions, due to inverter current limiting action, and are very difficult to detect through conventional means.

For addressing the aforementioned grid modeling problems, a detailed grid model (referred to herein as the CTM) was proposed by the authors in a companion paper [20], which has sufficient detail of the current flows on the power line from each individual DER connected to the grid. The main objectives of this paper are as follows. First, for the first time, it is shown how the accuracy of implementing AI algorithms on such a detailed grid model can be improved. Specifically, implementing SVM algorithm on the proposed CTM model is explored. Second, in addition to using fault current flows as the only input feature, the "traced" current information to expand the dimension of the feature space is exploited. That is, the traced current is used along with the power line current as the expanded feature space to identify the fault current which is a departure from existing methods. Finally, the performance of the combination of CTM with the SVM is demonstrated in the practical scenario of fault identification with DERs operating in distribution grids. Specifically, this work shows how applying this hybrid method can improve sensitivity in detecting very low level faults.

This paper is organized as follows. In Section 2, the weaknesses of using traditional grid models for fault current identification are pointed out, and applying CTM for multiple current sources of a distribution grid is proposed. The SVM method is implemented for fault current identification using the traced current in Section 3. Simulation results are presented in Section 4 and are discussed in Section 5, with concluding remarks given in Section 6.

2. Proposed Tracing Method

2.1. Single Power Line Fault Current Threshold

Consider the case where a DER group consists for example of solar and wind generation, a vehicle-to-grid support battery storage, and loads that are attached to bus 1 are connected to the end of the power distribution grid through a single power line as shown in Figure 1.

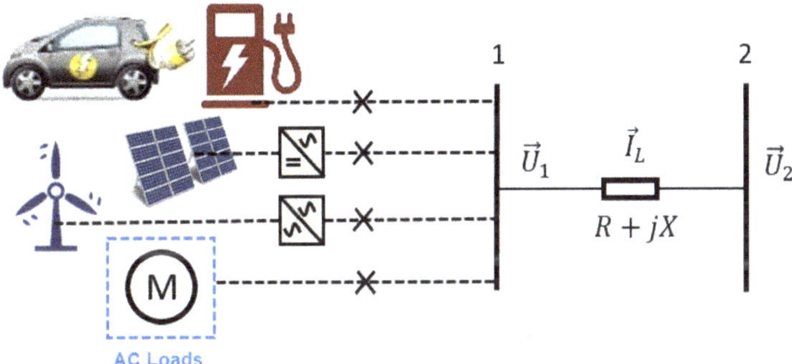

Figure 1. Single power line with distributed energy source (DER) customers.

In Figure 1, $\vec{I}_L$ represents the power line current, and $\vec{U}_1$ and $\vec{U}_2$ are the voltages on buses 1 and 2, respectively. The applied voltage difference angle of $\vec{U}_1 - \vec{U}_2$ is ψ. Without the impact of the DER group, $\vec{I}_L$ flows from bus 2 to bus 1 and the fault current can be easily identified by setting the fault current threshold. However, when the impact of renewable energy sources is considered, as in this case, the fault current threshold is difficult to determine, as the DERs may contribute to the fault current and the detection threshold varies. The power line's current value and direction may also vary with the impact of DERs, which complicates the discrimination of faults from the normal condition. It is therefore necessary to know significantly more details of the current to determine if there is a fault on the power line or at the load.

2.2. Current Tracing

Using Figure 1 to demonstrate the CTM deduction, without loss of generality, it is assumed that the power line impedance is

$$Z = R + jX, \tag{1}$$

where $R > 0$ and $X > 0$ are the resistance and reactance of the power line, respectively. An alternative case, when $X < 0$, is discussed in the companion conference paper [20]. In this paper, only the $X > 0$ case is considered. The impedance angle is θ. Based on Kirchhoff's current law, it follows that

$$\vec{I}_L = \sum I_i e^{j\phi_i}. \tag{2}$$

where $\vec{I}_i = I_i e^{j\phi_i}$ is the current of the ith DER, and $I_i > 0$ and ϕ_i are the magnitude and phase angle of $\vec{I}_i$, respectively.

Figure 1 can be expressed with an equivalent circuit as shown in Figure 2a. The equivalence holds as shown in Equations (3) and (4):

$$R_E = \frac{R^2 + X^2}{R}, \tag{3}$$

$$X_E = \frac{R^2 + X^2}{X}, \tag{4}$$

where $R_E > 0$ and $X_E > 0$ are the equivalent resistance and reactance respectively. Naturally, the equivalent circuit has the same total resistance, current, and power of the original circuit.

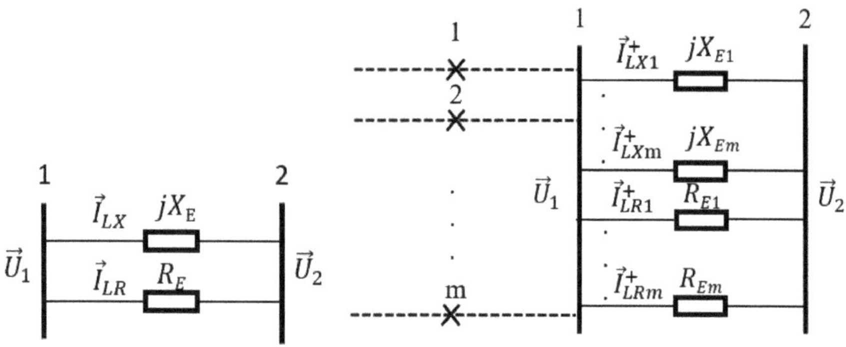

(a) Active-reactive current   (b) Multiple current sources tracing

Figure 2. Equivalent circuit.

In the equivalent circuit, the power line current $\vec{I}_L$ is virtually split into two parts defined as active current $\vec{I}_{LR}$ and reactive current $\vec{I}_{LX}$, which flows through R_E and X_E, respectively:

$$\vec{I}_L = \vec{I}_{LR} + \vec{I}_{LX}, \tag{5}$$

$$\vec{I}_{LR} = I_{LR}e^{j\psi}, \tag{6}$$

$$I_{LR} = I_L cos(\theta), \tag{7}$$

$$\vec{I}_{LX} = I_{LX}e^{j(\psi-\frac{\pi}{2})}, \tag{8}$$

$$I_{LX} = I_L sin(\theta), \tag{9}$$

where I_{LR} and I_{LX} are the magnitude of the active current and reactive current, respectively. Likewise, all current sources attached to bus 1 follow the same rule:

$$\vec{I}_i = \vec{I}_{Ri} + \vec{I}_{Xi} = I_{Ri}e^{j\psi} + I_{Xi}e^{j(\psi-\frac{\pi}{2})}, \tag{10}$$

$$I_{Ri} = I_{Li}cos(\psi - \phi_i), \tag{11}$$

$$I_{Xi} = I_{Li}sin(\psi - \phi_i), \tag{12}$$

where I_{Ri} and I_{Xi} are the magnitude of the active and reactive current from $\vec{I}_i$, respectively, which could either be positive or negative. From Kirchhoff's current law,

$$\sum_{I_{Ri}>0} (\vec{I}_{Ri}) + \sum_{I_{Ri}<0} (\vec{I}_{Ri}) = \sum (\vec{I}_{Ri}) = \vec{I}_{LR}, \tag{13}$$

$$\sum_{I_{Xi}>0} (\vec{I}_{Xi}) + \sum_{I_{Xi}<0} (\vec{I}_{Xi}) = \sum (\vec{I}_{Xi}) = \vec{I}_{LX}, \tag{14}$$

where $\sum_{I_{Ri}>0} (\vec{I}_{Ri})$ and $\sum_{I_{Ri}<0} (\vec{I}_{Ri})$ stand for the positive and negative part of $\vec{I}_{LR}$, respectively, and $\sum_{I_{Xi}>0} (\vec{I}_{Xi})$ and $\sum_{I_{Xi}<0} (\vec{I}_{Xi})$ represent the positive and negative part of $\vec{I}_{LX}$, respectively. The positive part is responsible for supplying the load as well as feeding extra current to the power distribution grid.

The negative part is responsible for absorbing current from positive part. The jth positive current source flowing through the power line is

$$\vec{I}^{+}_{LRj} = I^{+}_{LRj}e^{j\psi},\tag{15}$$

$$\vec{I}^{+}_{LXj} = I^{+}_{LXj}e^{j\left(\psi-\frac{\pi}{2}\right)},\tag{16}$$

$$I^{+}_{LRj} = \frac{I_{LR}I_{Rj}}{\sum\limits_{I_{Ri}>0}\left(I_{Ri}\right)},\tag{17}$$

$$I^{+}_{LXj} = \frac{I_{LX}I_{Xj}}{\sum\limits_{I_{Xi}>0}\left(I_{Xi}\right)}.\tag{18}$$

where $\vec{I}^{+}_{LRj}$ is the active part of the jth positive current source flowing through the power line with magnitude of I^{+}_{LRj}, and $\vec{I}^{+}_{LXj}$ is the reactive part of the jth positive current source flowing through the power line with magnitude of I^{+}_{LXj}. Therefore, the distribution grid model where each of the currents are independent from one another can be shown in Figure 2b.

Combining the active and reactive components of the jth positive current source flowing through the power line, the equivalent circuit with part of the jth positive current source that flows through the power line can also be derived such that $\vec{I}^{+}_{LZj} = \vec{I}^{+}_{LRj} + \vec{I}^{+}_{LXj}$, with its impedance, Z_{Ej}, as shown in Figure 3. The complete derivation of this situation can be found in the companion paper [20].

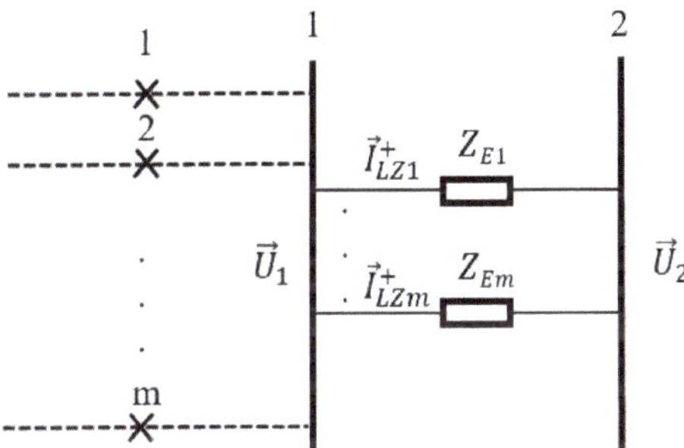

Figure 3. Equivalent circuit of impedance lines.

3. Support Vector Machine and Current Tracing Kernel

3.1. Binary Classification Problem Formation

Given a set of power line current measurements $Z = \{z_i, i = 1...n\}$, $z_i \in \mathbb{R}^m$ that may or may not contain fault current and the set of labels $Y = \{y_i, i = 1...n\}$, $y_i \in \{0,1\}$, m stands for the dimension of the measurement, and n is the number of observations. The fault current identification problem can be modeled as a binary classification problem by establishing the connection between the above two sets such that

$$y_i = \begin{cases} -1 & \text{if } a_i = 0, \\ 1 & \text{if } a_i \neq 0. \end{cases}\tag{19}$$

$y_i = 1$ indicates that the ith current measurement is fault current, or, alternatively, there are no fault currents for $y_i = -1$.

3.2. Support Vector Classifiers

The classification problem is reformatted into the optimization problem in [21]. The objective function can be defined as

$$\min \ \frac{1}{2}||\omega||^2 + C\sum_{i=1}^{m}\xi_i,\tag{20}$$

where C is used to control the penalty of the misclassification, ω is a constant such that $\omega = [\omega_0, \omega_1, ...\omega_m]^T$, and ξ_i is the slack variable. The objective function is subject to the constraint that

$$\omega_0 + \omega^T z_i \geq 1 - \xi_i \ if \ y_i = 1,\tag{21}$$

$$\omega_0 + \omega^T z_i \leq \xi_i - 1 \ if \ y_i = -1.\tag{22}$$

3.3. Support Vector Machines

As an extension of the support vector classifier, SVM is established by enlarging the feature space using kernel. Kernel is a function that is used to quantify the similarity of two observations. A linear Kernel is defined as the inner product of two observations [21]:

$$K(z_i, z_{i'}) = \sum_{j=1}^{m} z_{i,j} z_{i',j},\tag{23}$$

where z_i and $z_{i'}$ are the two observations, and $z_{i,j}$ and $z_{i',j}$ are the observations on jth dimension.

In addition, some commonly used kernels are [21]

- Polynomial kernel: $K(z_i, z_{i'}) = (1 + \sum_{j=1}^{m} z_{i,j} z_{i',j})^d$
- Radial kernel: $K(z_i, z_{i'}) = exp(-\gamma \sum_{m}^{j=1}(x_{i,j} - x_{i',j})^2)$

where γ and d are positive constants and r is a constant.

3.4. Current Tracing Kernel

In [22], the author applied principle component analysis to select the best features with highest information content to identify faults. This study concluded that the top three features for fault identification were reactive power, real power, and angle of voltage. Interestingly, current is not one of them. The author concluded that with all of the three selected features, the accuracy of identification is almost 96%. With all of the six features, i.e., the above features plus magnitude and angle of current and magnitude of voltage, the accuracy of identification is no more than 97%. In the proposed approach, current is used as the only feature. Without current tracing, the feature space consists of only the line current magnitude and angle.

Based on Equations (15)–(18), the line current can be decomposed into several traced currents flowing through virtual impedance lines as shown in Figure 3. The feature space of line currents can be enlarged by using the traced currents:

$$K(I_L, e^{j(\psi-\theta)}) = \vec{I}_{LZj}^{+}.\tag{24}$$

where K represents the linear mapping from power line current to the traced current.

4. Simulation Results

4.1. Current Tracing Kernel Results

In this simulation, the proposed CTM is applied to the single line system of Figure 4. Both sides of the single line have a group of DERs and loads, and bus 2 is connected directly to the external distribution grid. All of the loads are of the constant power type, and the DERs are static generators.

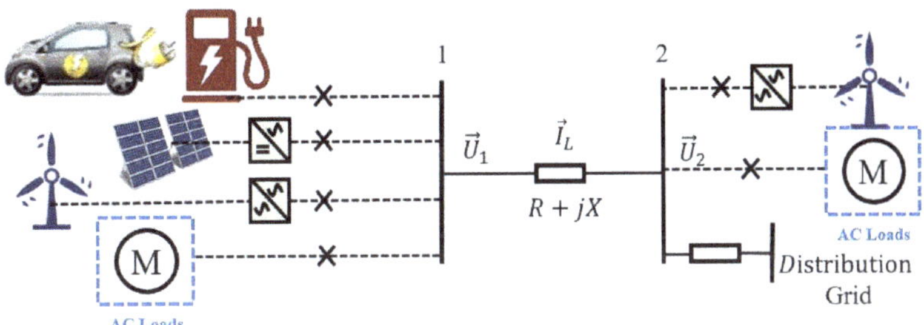

Figure 4. Multiple source to multiple source on single power line.

The current sources parameters are listed in Table 1, which are given in the format of active and reactive power. DERs that connect to bus 1 cannot support the AC loads attached to bus 1 so that the currents flow from bus 2 to bus 1, which is opposite from the case shown in the companion paper [20]. Bus 2 is selected as the reference bus and the single power line is 20 km length with series impedance of 0.121 + j0.107 Ω/km.

Table 1. The current source parameters.

Bus	Label	Power	
		Active/MW	Reactive/MVar
	1	−20	−10
1	2	−20	−5
	3	30	15
	4	20	5
	1	−20	−10
2	2	40	15
	3	30.872544	10.771589

In Table 1, the positive elements indicate an absorption of power whereas negative elements represents generating power. The third current source attached to bus 2 is the external grid and is calculated by the power flow. Equations (15) and (16) are applied to find the traced current on the power line from each bus as shown in Table 2. All of the traced currents are listed in per unit value. The traced current magnitude and phase in radians are selected as the kernel for fault current identification.

Table 2. Current tracing results with Equations (15)–(18).

Bus	Label	Active Current Tracing		Reactive Current Tracing	
		Mag	Phase/°	Mag	Phase/°
	1	0	0	0	0
1	2	0	0	0	0
	3	5.3677	13.5251	5.6267	−76.4749
	4	3.8535	13.5251	2.5276	−76.4749
	1	0	0	0	0
2	2	3.5366	13.5251	3.6588	−76.4749
	3	5.6847	13.5251	4.4955	−76.4749

In Table 2, a zero value indicates that the corresponding current source does not contribute to the current on the power line. It can be observed that the zero value occurs at bus 1; labels 1 and 2; and bus 2, label 1. This does not violate common sense as these are all labeled as loads that are consuming active and reactive power and do not contribute to the power line current. Moreover, all of the traced active currents have the same phase angle regardless of how the current is traced from bus 1 or bus 2 in either direction. The same rationale applies to the reactive currents. This is also consistent with the fact that the voltages applied on the buses do not change when the current is decomposed into its traced components. The reactive current is 90 degrees out of phase from the active current, which complies with Equations (15) and (16).

It is observed that, if all of the traced currents from the same bus are summed, the result is equivalent to the total power line current. This proves the equivalence of the current tracing theory as the current tracing will not lose or generate new currents in addition to the line current.

4.2. SVM Results

In distribution systems with DERs, the fault current can be very small, as inverter-based DERs can only produce exceptionally small fault current contributions due to inverter current limiting action. Moreover, injected currents on different loads are continuously fluctuating in the normal condition. To obtain representative currents in the single line system, sample noises are injected to the specified load powers in Table 1 and the power flow is recalculated to obtain the traced current. This process is then repeated to obtain the load profile and a continuous currents curve. The injected noise follows the normal distribution such that,

$$X \sim N(\mu, \sigma^2),\tag{25}$$

where X represents active or reactive power sample noises; μ represents the average of the sample noises, which is set to 0; and σ stands for the standard deviation, which is set to 0.1. All the sample noises are independent from each other. The sample noises were injected cumulatively to the loads such that the kth point on the load profile is

$$P_k = P + \sum_{i=1}^{k} Xp_i,\tag{26}$$

$$Q_k = Q + \sum_{i=1}^{k} Xq_i,\tag{27}$$

where P_k and Q_k represent the active and reactive power of the kth point of the load profile, respectively; P and Q represent the given active and reactive power (Table 1); and Xp_i and Xq_i stand for the ith active or reactive power sample noises, respectively.

This process is repeated 500 times, and all of the parameters that are used for current tracing and SVM training purpose are recorded. In addition to sample noises, a small fault is also injected by increasing the active power consumption of bus 1, with current source 3 increased by 10% and decreasing the reactive power by 10% of the same current source. Again, the process is repeated 500 times, and all parameters are recorded. Only the traced current on the power line from bus 2 side is used as the current tracing kernel; however, it is the same as if the traced current from bus 1 side was used as the current tracing kernel, as in the conference paper [20]. The first 500 parameters are taken as normal condition, i.e., $y_i = -1$, and the last 500 parameters are considered as fault condition, i.e., $y_i = 1$. The penalty of misclassification C is set to be 1. Seventy percent of the parameters are randomly selected as the training data, and the remaining 30% are taken as the testing data. In this work, the non-waveform phasor current information is used for the fault identification problem. Other alternative measurement data have been considered in other research, such as exploiting sub-cycle waveform distortion features in pattern recognition algorithms as a part of the identification process [12,15].

The confusion matrix is used to represent the testing results as defined in [23]. The f-score, recall, and precision parameters are used to evaluate the performance of fault current identification based on the confusion matrix such that,

$$prec = \frac{tp}{tp + fp}, \tag{28}$$

$$rec = \frac{tp}{tp + fn}, \tag{29}$$

$$fs = 2\frac{prec * rec}{prec + rec}, \tag{30}$$

where tp, fp, tn, and fn represent true positive, false positive, true negative, and false negative, respectively.

To show the advantage of using the current tracing kernel, the fault current identification results using different feature spaces is compared. First, only the power line current $\vec{I}_L$ is used as the feature space. Then, the polynomial kernel, radial kernel, and current tracing kernel are added as the expanded feature space. All of the confusion matrices are calculated based on the same training and testing data. The confusion matrices and the performance based on the confusion matrices are shown in Tables 3 and 4, respectively.

Table 3. Confusion matrix using different feature space.

Feature Space		Fault	Normal
No Kernel	Predict Fault	tp = 90	fp = 61
	Predict Normal	fn = 13	tn = 136
Polynomial Kernel	Predict Fault	tp = 100	fp = 51
	Predict Normal	fn = 19	tn = 130
Radial Kernel	Predict Fault	tp = 95	fp = 56
	Predict Normal	fn = 16	tn = 133
Current Tracing Kernel	Predict Fault	tp = 145	fp = 6
	Predict Normal	fn = 0	tn = 149

Table 4. Performance using different feature space.

Feature Space	Precision	Recall	f1-Score
No Kernel	0.596	0.874	0.709
Polynomial Kernel	0.662	0.840	0.74
Radial Kernel	0.629	0.856	0.725
Current Tracing Kernel	0.96	1	0.98

It is clearly seen that among all the feature spaces used, the current tracing kernel has the best performance. All three performance parameters are significantly higher than the other feature spaces that were used. The recall value equals 1, which indicates that when there is a fault current on the power line; the SVM method using current tracing kernel will definitely detect it. In addition, the polynomial kernel and radial kernel have a better performance than if only $\vec{I}_L$ is used as feature space. However, the performance parameters do not increase significantly. When comparing with the results shown in [22], which have 97% overall accuracy, the overall current tracing kernel result in this paper has improved, i.e., an f1-score value of 98%.

5. Discussion

In the companion paper, it has been verified that the proposed tracing method is mathematically and physically identical to the original network. In this paper, the developed technique has been applied to a simple single power line system; however, it could be generalized to be as a part of a larger distribution grid. Using the proposed CTM, pipelines of each current sources' contribution towards the fault currents were established in the form of virtually decomposed traced currents. The expanded features of the traced currents have the distinct advantage of containing more detailed information of the fault current, such as the change in contribution of each current source and the phase shift caused by the fault. These features can be exploited by the SVM algorithm for improved classification of faults, particularly where the fault levels are very low from the current limiting action of inverter-based DERs.

By choosing a large sample set of 1000 data points for faulted and normal cases, the proposed SVM method combined with the current tracing kernel is shown to be much more sensitive to very low-level faults on the power line compared to the polynomial and radial kernel methods, where the accuracy of detection was at most 74%. Compared to the other methods, such as the one in [22] where the accuracy of detection was at most 97%, the proposed method has improved accuracy of 98% while using significantly less measurement features.

These results are relevant in the context of smart grids aiming to improve distribution system state estimation, extending supervisory control and data acquisition processes beyond the substation domain as DERs proliferate. In these simulations, only the batch data is used for training and testing. However, the proposed model can be easily applied to practical streaming data obtained from intelligent electronic devices used in advanced metering infrastructure such as protection relays and phasor measurement units. With the enhanced features extracted through the proposed CTM, more intelligent protection relay coordination and fault isolation may be possible, particularly when considering multiple inverter-based DER operation from energy storage and renewable energy.

6. Concluding Remarks

In this paper, we propose a CTM augmented with SVM method to model and test a distribution feeder for power line fault current identification. The CTM modeled the distribution network by providing a detailed map of how current flows from each current source that is connected to one bus towards another source tied to a different bus. The proposed method does not violate any physical circuit laws. After applying the proposed current tracing, the virtual traced currents and the corresponding circuit are exactly equivalent to the original current and circuit. The traced current provides sufficient details and sensitivity for identifying faults and abnormal conditions in the distribution feeder. With these details, the feature space of the power line current is enlarged through the "current tracing kernel". In addition, the results proved and demonstrated the proposed method on a single power line distribution system, and the SVM method's performance was evaluated and compared by using different kernel methods. The results indicate that with the benefits of the proposed current tracing kernel, the SVM method is enhanced with more sensitivity to very low level faults compared to the commonly used kernel such as polynomial kernel and radial kernel.

The proposed method is a good fit for the distribution grid primary side overcurrent protection scheme. In the companion paper, the multicurrent sources to multicurrent source current tracing are already introduced, and they can be used to further expand the feature spaces of fault current. In the future, the authors will explore the implementation aspects of the proposed CTM described herein with the distribution grid backup protection scheme, especially the multicurrent sources to multicurrent sources case, which would undoubtedly occur with higher DER penetration in the future. Furthermore, in future work, the authors will explore how to implement and test this approach on a laboratory scale distribution test feeder to further verify the implementation of the SVM detection scheme using the proposed tracing method.

Author Contributions: Conceptualization, W.F.; methodology, W.F. and P.M.; software, W.F.; validation, W.F. and P.M.; formal analysis, W.F.; investigation, W.F.; resources, W.F.; data curation, W.F.; writing–original draft preparation, W.F.; writing–review and editing, P.M.; visualization, W.F.; supervision, P.M.; project administration, P.M.; funding acquisition, P.M.

Funding: This research was funded in part by the Oklahoma Center for the Advancement of Science and Technology (Project No. AR18-073) and the Oklahoma Gas & Electric Company (Project No. A18-0274).

Conflicts of Interest: The authors declare no conflicts of interest.

Abbreviations

The following abbreviations are used in this manuscript.

AI   Artificial Intelligence
CTM   Current Tracing Method
DER   Distributed Energy Resource
KNN   K-nearest Neighbors
SVM   Support Vector Machine

References

1. Saad, H.; Peralta, J.; Dennetiere, S.; Mahseredjian, J.; Jatskevich, J.; Martinez, J.; Davoudi, A.; Saeedifard, M.; Sood, V.; Wang, X.; et al. Dynamic averaged and simplified models for MMC-based HVDC transmission systems. *IEEE Trans. Power Deliv.* **2013**, *28*, 1723–1730. [CrossRef]
2. Daryabak, M.; Filizadeh, S.; Jatskevich, J.; Davoudi, A.; Saeedifard, M.; Sood, V.; Martinez, J.; Aliprantis, D.; Cano, J.; Mehrizi-Sani, A. Modeling of LCC-HVDC systems using dynamic phasors. *IEEE Trans. Power Deliv.* **2014**, *29*, 1989–1998. [CrossRef]
3. Runolfsson, T. On the dynamics of three phase electrical energy systems. In Proceedings of the IEEE American Control Conference (ACC), Boston, MA, USA, 6–8 July 2016; pp. 6827–6832.
4. Aravinthan, V.; Balachandran, T.; Ben-Idris, M.; Fei, W.; Heidari-Kapourchali, M.; Hettiarachchige-Don, A.; Jiang, J.N.; Lei, H.; Liu, C.C.; Mitra, J.; et al. Reliability modeling considerations for emerging cyber-physical power systems. In Proceedings of the 2018 IEEE International Conference on Probabilistic Methods Applied to Power Systems (PMAPS), Boise, ID, USA, 24–28 June 2018; pp. 324–330.
5. Davis, K.R.; Davis, C.M.; Zonouz, S.A.; Bobba, R.B.; Berthier, R.; Garcia, L.; Sauer, P.W. A cyber-physical modeling and assessment framework for power grid infrastructures. *IEEE Trans. Smart Grid* **2015**, *6*, 2464–2475. [CrossRef]
6. Fei, W.; Jiang, J.N.; Wu, D. Impacts of Modeling Errors and Randomness on Topology Identification of Electric Distribution Network. In Proceedings of the 2018 IEEE International Conference on Probabilistic Methods Applied to Power Systems (PMAPS), Boise, ID, USA, 24–28 June 2018; pp. 1–6.
7. Ji, G.; Sharma, D.; Fei, W.; Wu, D.; Jiang, J.N. A Graph-theoretic Method for Identification of Electric Power Distribution System Topology. In Proceedings of the 1st Global Power, Energy and Communication Conference (GPECOM), Nevsehir, Turkey, 12–15 June 2019; pp. 403–407.
8. Garces, A. A linear three-phase load flow for power distribution systems. *IEEE Trans. Power Syst.* **2016**, *31*, 827–828. [CrossRef]
9. Yang, J.; Zhang, N.; Kang, C.; Xia, Q. A state-independent linear power flow model with accurate estimation of voltage magnitude. *IEEE Trans. Power Syst.* **2017**, *32*, 3607–3617. [CrossRef]
10. Vita, V.; Ekonomou, L.; Christodoulou, C.A. The impact of distributed generation to the lightning protection of modern distribution lines. *Energy Syst.* **2016**, *7*, 357–364. [CrossRef]
11. Fei, W.; Ji, G.; Sharma, D.; Jiang, J.N. A New Traveling Wave Representation for Propagation of Energy Transients in Power Lines from a Quantum Perspective. In Proceedings of the North American Power Symposium (NAPS), Fargo, ND, USA, 9–11 September 2018; pp. 1–6.
12. Pavlatos, C.; Vita, V. Linguistic representation of power system signals. In *Electricity Distribution*; Springer: Berlin, Germany, 2016; pp. 285–295.
13. Harrou, F.; Taghezouit, B.; Sun, Y. Improved *k* NN-Based Monitoring Schemes for Detecting Faults in PV Systems. *IEEE J. Photovolt.* **2019**, *9*, 811–821. [CrossRef]
14. Abid, F.B.; Zgarni, S.; Braham, A. Distinct bearing faults detection in induction motor by a hybrid optimized SWPT and aiNet-DAG SVM. *IEEE Trans. Energy Convers.* **2018**, *33*, 1692–1699. [CrossRef]
15. Pavlatos, C.; Vita, V.; Dimopoulos, A.C.; Ekonomou, L. Transmission lines' fault detection using syntactic pattern recognition. *Energy Syst.* **2019**, *10*, 299–320. [CrossRef]

16. Guo, M.F.; Zeng, X.D.; Chen, D.Y.; Yang, N.C. Deep-learning-based earth fault detection using continuous wavelet transform and convolutional neural network in resonant grounding distribution systems. *IEEE Sens. J.* **2017**, *18*, 1291–1300. [CrossRef]

17. Park, J.D.; Candelaria, J.; Ma, L.; Dunn, K. DC ring-bus microgrid fault protection and identification of fault location. *IEEE Trans. Power Deliv.* **2013**, *28*, 2574–2584. [CrossRef]

18. Vaseghi, B.; Takorabet, N.; Meibody-Tabar, F. Fault analysis and parameter identification of permanent-magnet motors by the finite-element method. *IEEE Trans. Magn.* **2009**, *45*, 3290–3295. [CrossRef]

19. Cheng, F.; Peng, Y.; Qu, L.; Qiao, W. Current-based fault detection and identification for wind turbine drivetrain gearboxes. *IEEE Trans. Ind. Appl.* **2016**, *53*, 878–887. [CrossRef]

20. Fei, W.; Moses, P. Modeling Power Distribution Grids through Current Tracing Method. In Proceedings of the IEEE International Conference on Smart Energy Grid Engineering (SEGE), Oshawa, ON, Canada, 12–14 August 2019; pp. 196–200.

21. James, G.; Witten, D.; Hastie, T.; Tibshirani, R. *An Introduction to Statistical Learning*; Springer: Berlin, Germany, 2013; Volume 112.

22. Zhang, Y.; Ilic, M.D.; Tonguz, O.K. Mitigating blackouts via smart relays: A machine learning approach. *Proc. IEEE* **2010**, *99*, 94–118. [CrossRef]

23. Sammut, C.; Webb, G.I. *Encyclopedia of Machine Learning and Data Mining*; Springer Publishing Company, Incorporated: Berlin, Germany, 2017.

Article

Comparison and Design of Resonant Network Considering the Characteristics of a Plasma Generator

Geun Wan Koo [1], Won-Young Sung [2] and Byoung Kuk Lee [1,*]

[1] Department of Electrical and Computer Engineering, Sungkyunkwan University, 2066, Seobu-ro, Jangan-gu, Suwon-si 16419, Gyeonggi-do, Korea

[2] Automotive Research & Development Division, Hyundai Motor Group, 150, Hyundaiyeonguso-ro, Namyang-eup, Hwaseong-si 18280, Gyeonggi-do, Korea

* Correspondence: bkleeskku@skku.edu; Tel.: +82-31-299-4612

Received: 22 July 2019; Accepted: 15 August 2019; Published: 16 August 2019

Abstract: This paper presents a theoretical analysis and experimental study on the resonant network of the power conditioning system (PCS) for a plasma generator. In order to consider the characteristics of the plasma load, the resonant network of the DC-AC inverter is designed and analyzed. Specifically, the design of an LCL resonant network and an LCCL resonant network, which can satisfy the output current specification in consideration of plasma characteristics, is explained in detail. Moreover, the inverter current and phase angle between the inverter voltage and current is derived for evaluating inverter performance. Based on these analysis results, the DC-AC inverter can be designed for a plasma generator considering plasma load characteristics. The theoretical analysis of both networks is validated through the simulation and experimental results.

Keywords: plasma generator; LCL network; LCCL network; phase compensation; ZVS control

1. Introduction

Plasma generators are used in various industrial fields for display panels and in the wafer cleaning process of semiconductors. According to the growth in the display and semiconductor markets, plasma generators are becoming increasingly important in industrial fields [1,2]. These plasma generators are conventionally constructed as a power conditioning system (PCS) and a chamber for plasma generation. High frequency current must be supplied by the PCS to the reactor, which is a magnetic substance in the chamber for supply energy which is used to ionize the gas entering the chamber. The electric field for ionizing is generated according to the frequency, sinusoidal wave, and magnitude of the supplied current. Therefore, in order to effectively generate the electric field, the PCS should supply high switching frequency and low total harmonic distortion (THD) constant current [3,4].

Another characteristic of the PCS for a plasma generator is that the inverter of the PCS using zero voltage switching (ZVS) is conventionally adopted to decrease the switching losses. In order to implement ZVS, the phase shift technique or load resonant technique has been used in previous research on the inverter. In order to achieve ZVS, the inverter should be designed in consideration of the following load characteristics in the plasma load case [5]: (1) the resistance of the load is increased in proportion to the gas injected into the chamber; (2) the resistance of the load is changed in inverse proportion to the load current [6]; (3) a constant load current is recommended for maintaining the stable plasma state; (4) the THD of the load current should be low for effective plasma generation [7–9].

In the case of the conventional resistive load, the pulse frequency modulation (PFM) or phase shift control are applied for the ZVS operation of the inverter. In plasma load, applying the PFM to the control output current is difficult due to the characteristics of the plasma load mentioned above. When the PFM is used to vary the output current in the plasma load, the plasma load resistance is varied depending on the output current. If the output current for generating plasma is increased, the resistance of the plasma load is decreased. When the output current for generating plasma is decreased, the resistance of the plasma load is decreased in an inversely proportional manner [6]. In addition, the designed initial Q-factor is varied rapidly due to changes in resistance. Because of the changed Q-factor, the resonant inverter cannot control the output current to the designed value.

Therefore, regulating the inverter output current using PFM is difficult because the plasma load resistance is changed again and the phenomena mentioned previously occur repeatedly. In order to solve these problems, the phase shift control should be used for the output current regulation. Conventionally, the phase shift full bridge inverter and load resonant inverter can solve the problem of using the phase shift control. In the plasma load case, the harmonics of the output current should be strictly regulated in order to satisfy plasma quality as characteristics of the plasma load [10–13]. Hence, it is difficult to adopt a phase shift full bridge inverter for the plasma system, and only the load resonant inverter using the phase shift method can be adopted. If a phase shift inverter is used for the plasma load, the filter with the ability to implement the sinewave of the load current is required for designing a phase shift inverter [14,15]. Using the additional filter on the phase shift inverter leads to increased system cost and volume. For this reason, the load resonant inverter using phase shift control is preferred for the plasma load system over the phase shift inverter with an additional filter [15].

In order to adopt a load resonant inverter in the plasma system, the effective resonant network design should be considered. There are two representative resonant networks for plasma systems: the structure of an LCL resonant network and the structure of an LCCL resonant network. These networks have different characteristics of the maximum inductor current and phase angle between inverter voltage ($v_{o.inv}$) and inverter current ($i_{o.inv}$). In the resonant network design, the maximum inductor current and inverter phase are the factors that influence the conduction loss and the soft-switching range for phase shift control. Therefore, in order to design a resonant network that is suitable for a plasma generator, the characteristics of the LCL network and the LCCL network should be compared to select the optimal maximum inductor current and phase angle.

In this paper, in order to explain the specialty of plasma load to the design of the resonant network, both the LCL network and LCCL network, which are conventionally used for resonant inverter systems and satisfy the specification of the plasma load, are analyzed in detail. Based on the analysis, the LCL resonant network and LCCL network are designed in consideration of the characteristics of the plasma load and with the aim of preventing the drop-out phenomenon. In order to evaluate the designed networks, simulation and analysis are conducted. Finally, the experimental results based on a 1 kW plasma generator are presented to verify the performances of the LCCL and LCL resonant networks on the plasma generator.

2. Control Scheme and Characteristics of the Plasma Load

Figure 1 shows the conceptual circuit diagram of the plasma generation system that generates plasma by injecting gas into the reactor of the chamber. In Figure 1, L_r, C_r, L_{lkg}, and R_{plasma} represent the resonant inductor, resonant capacitor, leakage inductor, and the equivalent resistance of the plasma load, respectively. The "V_{in}" represents the input voltage of the inverter. $v_{o.inv}$, $i_{o.inv}$, and i_{plasma} indicate the mean output voltage of the inverter, the output current of the inverter and the output current for generating plasma, respectively. In the plasma generation system, the injected gas affects the load. This plasma generator has unique features: (1) the impedance of the plasma load is proportionally decreased according to the increase in i_{plasma} value [6] and (2) the sinusoidal wave is highly recommended for generating plasma [15]. These characteristics are an important consideration point for designing the power supply of the plasma system.

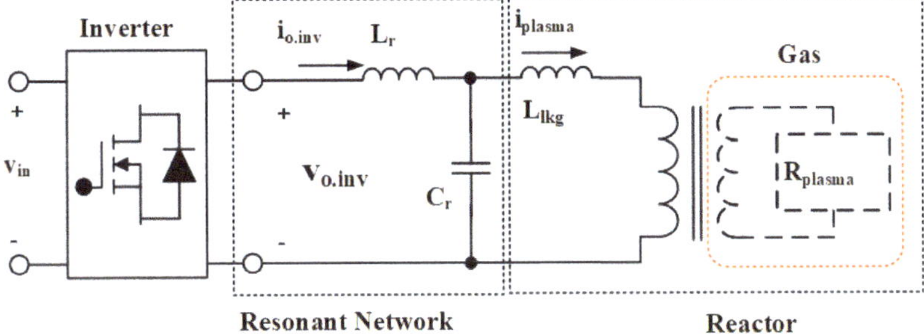

Figure 1. Conceptual circuit diagram of the plasma generation system.

2.1. Chracteristics Analysis of the Plasma Load

These characteristics cause a special issue in the plasma system. Among the above-mentioned plasma characteristics, the impedance of plasma load leads to a special issue when pulse frequency modulation is adopted for output current control. In Figure 2, in order to regulate the output current, the switching frequency (f_{sw}) is changed to be above resonant frequency (f_r). In the conventional gain curve of the resonant network shown in Figure 2, in order to regulate the output current, increasing or decreasing switching frequency is normally used in the resistive load.

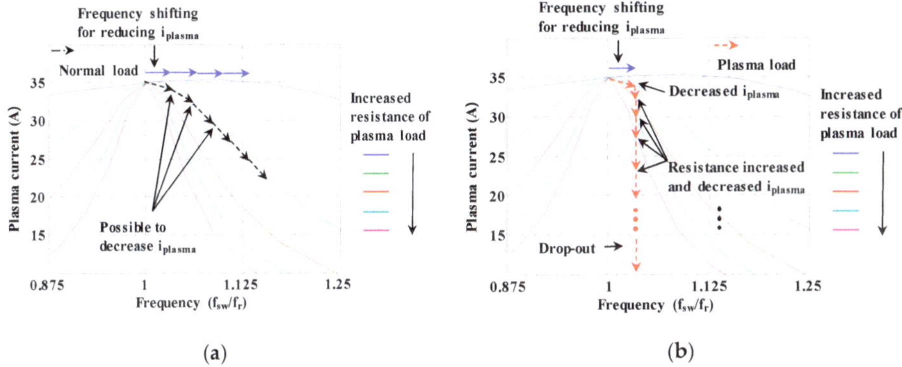

Figure 2. Description diagram of plasma drop-out: (**a**) frequency control method in resistive load; (**b**) frequency control method in plasma load.

However, in the plasma load case, when increasing the switching frequency to reduce the output current the plasma load impedance is increased according to the reduced output current. The increased impedance changes the Q-factor of the network to make a sharp current gain curve. In addition, the output current is decreased again because of the change in the current output gain curve, as shown in Figure 2. These mechanisms progress repeatedly until the output current reaches zero. Therefore, using PFM to control the output current is difficult in the plasma load. This phenomenon is called drop-out and is observed when controlling output current using PFM in the plasma load. In order to solve the above-mentioned problems, the phase shift control scheme should be adopted for the resonant inverter to satisfy the output current control range. Moreover, the inverter should be operated at the resonant frequency to avoid plasma drop-out.

2.2. Control Method Considering Characteristics of the Plasma Load

In order to obtain various output current ranges with ZVS, the resonant network should be designed to consider the minimum phase shift angle which can control the minimum plasma current with ZVS operation. The minimum phase shift angle is calculated by Equation (1) [14], which represents the relation between the amplitude of the fundamental wave of the inverter output voltage ($v_{o.inv.1}$) and the phase shift angle.

Figure 3 shows the phase shift control scheme for output current control in the resonant inverter. In this scheme, the inverter is operated in order to reduce the fundamental wave of the inverter output voltage ($v_{o.inv.1}$) with increasing phase shift angle for reducing the width of the square wave (β). In this control scheme, the $v_{o.inv.1}$ can be calculated using Equation (1) and the phase angle between the inverter output voltage and output current is the crucial factor during ZVS to obtain various output currents. According to Equation (1), $v_{o.inv.1}$ can be controlled to adjust the phase shift angle at the fixed resonant frequency. The output current is obtained by the relationship between $v_{o.inv.1}$ calculated using Equation (1) and the designed resonant network gain. Therefore, the maximum inverter current and phase angle between $v_{o \cdot inv}$ and $i_{o \cdot inv}$ should be considered in designing the resonant network to achieve lower conduction loss and soft-switching.

$$v_{o.inv.1} = \frac{4V_{in} \cdot \sin\left(\frac{\beta}{2}\right)}{\pi} \cos\left(\omega t - \frac{\beta}{2}\right) \tag{1}$$

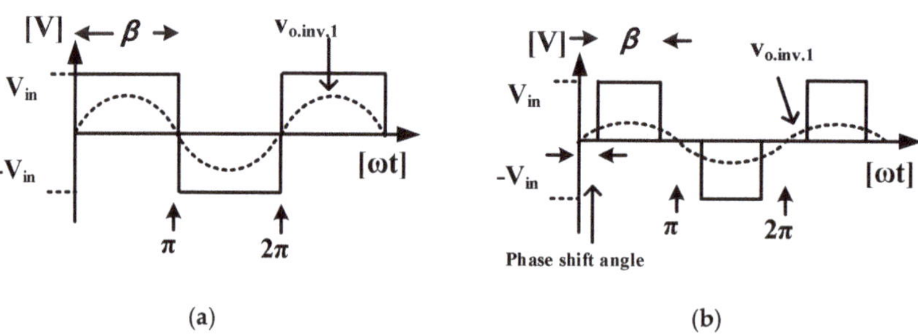

Figure 3. Conceptual diagram of control method considering plasma drop-out: (**a**) full output current condition and (**b**) decreasing output current.

3. Analysis of the Resonant Network for the Plasma Load

The LCL network and the LCCL network are suitable structures for the resonant network for plasma generation. Depending on the network, the phase angle between the inverter maximum current and the plasma current has different characteristics. Therefore, impedance analysis is necessary for considering the different characteristics of the resonant network depending on variations in the values of passive elements. Using the results of impedance analysis, the maximum current of the inverter, plasma current at the resonant frequency, and the phase between the inverter voltage and the current can be calculated.

3.1. Analysis of LCL Resonant Network

The LC resonant network is suitable for satisfying the constant current output of the plasma inverter with just a few passive elements. However, the LC resonant network is constructed here as the LCL network because of the leakage inductance of the reactor which is used for plasma generation, as shown in Figure 4. Therefore, in order to design the power supply for the plasma generator, the characteristics of the LCL network should be analyzed mathematically. The first consideration point in designing the network using mathematical analysis is deriving the L_r and C_r values, which can regulate the constant output current of the inverter regardless of load variation.

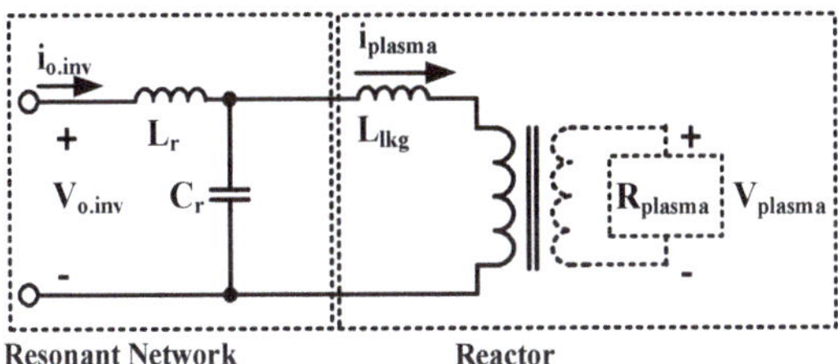

Figure 4. Conceptual circuit diagram of the LCL resonant network.

In order to regulate the constant output current while preventing drop-out, the resonant frequency should be selected as an operating frequency of the inverter, and the output current can be derived from impedance analysis, as shown in Equation (2) [14]. Figure 5a presents the characteristics of output current through the impedance analysis according to frequency variation. As shown in Figure 5a, the output current changes with the load variation, except for the resonant frequency. Therefore, the resonant network should be designed to satisfy the maximum output current value at the resonant frequency in the plasma generation system.

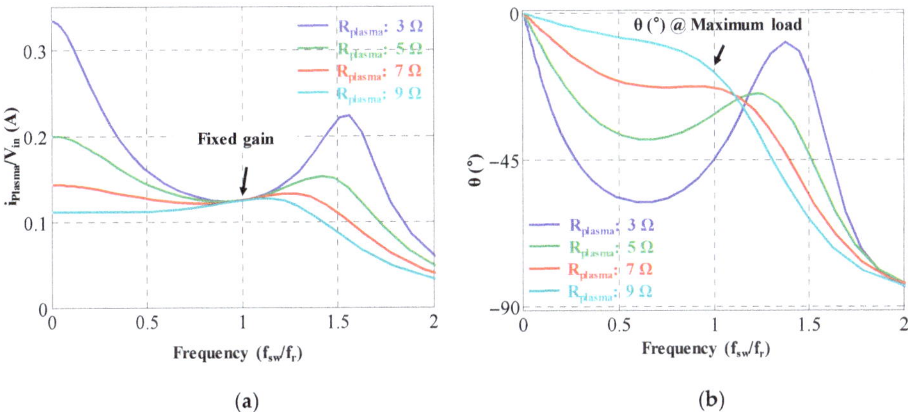

Figure 5. Characteristics of LCL network for designing a power conditioning system (PCS): (**a**) output current characteristics of the LCL network and (**b**) phase angle characteristics of the LCL network.

After determining the maximum current, phase shift control is necessary for satisfying the minimum current of the load requirement and preventing drop-out. The phase shift angle for satisfying the minimum output current with phase shift control can be calculated using Equation (1). The consideration point for satisfying the phase shift angle which is calculated to regulate the minimum output current in the inverter is the phase difference between the inverter voltage and the current. When the phase difference is smaller than the phase shift angle, the inverter is operated in the hard-switching region. By contrast, when the phase difference is larger than phase shift angle, the inverter could be operated in the soft-switching region. The relationship between the phase shift angle and the phase difference is considered using Equation (3) [16]. If the calculation results of Equation (3) are larger than the required phase shift angle, the inverter could be operated in the soft-switching region at resonant frequency. Figure 5b shows the phase difference between the inverter voltage and the current according to load variation. When the inverter is operated at the resonant frequency to prevent drop-out and when the resistance of the load is small, the phase difference is large. Furthermore, the phase difference becomes small when the resistance of the load is large, as shown in Figure 5b.

Using the above-mentioned analysis, the inverter can control the output current, which is required for plasma generation. Regarding the other point of consideration aside from controlling load current in order to design the power supply for the plasma generator, the inverter current ($i_{o \cdot inv}$) should be considered for the efficiency of the inverter. The inverter current, which is the main cause of inverter losses, can be calculated using Equation (4) [16]. This inverter current can be used for the index of the inverter efficiency. Therefore, in order to design a high efficiency inverter, the inverter current should be considered.

$$i_{plasma.LCL}(f_r) = \frac{V_{o.inv.1}\sqrt{C_r}}{\sqrt{L_r}} \tag{2}$$

$$\theta_{LCL}(f_r) = \tan^{-1}\left[\frac{(L_r - L_{lkg})}{R_{plasma}\sqrt{L_r C_r}}\right] \tag{3}$$

$$i_{o.inv.LCL}(f_r) = \sqrt{\frac{C_r}{L_r}}\frac{1}{L_r\sqrt{\frac{1}{(L_r - L_{lkg})^2 + C_r L_r R_{plasma}^2}}} \tag{4}$$

3.2. Analysis of LCCL Resonant Network

The other structure of the resonant network for applying the plasma inverter is the LCCL network, as shown in Figure 6. Using the LCCL network, it is possible to compensate for the phase difference, which is not possible with the LCL network. In the case of the LCCL network design, it is necessary to use impedance analysis, as shown in Equation (5), to regulate the constant output current regardless of the load at the resonant frequency in order to prevent drop-out. Figure 7a shows the maximum current of the LCCL network according to load variation. Aside from the resonant frequency, the output current is changed with the load variation, as shown in Figure 7a.

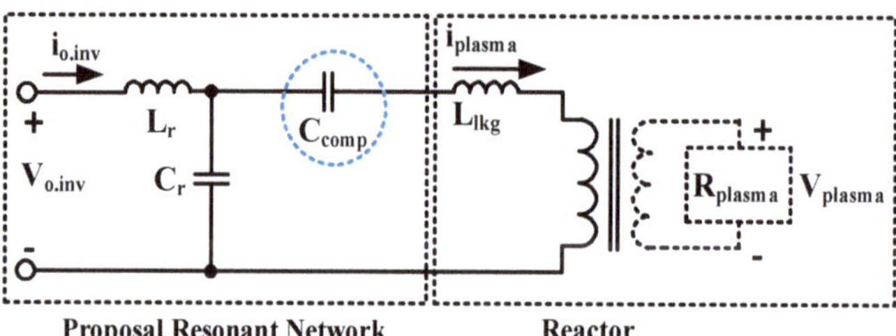

Figure 6. Conceptual circuit diagram of the LCCL resonant network.

The phase shift control is also necessary for satisfying the minimum output current of the LCCL network. In this case, the phase shift angle satisfying the minimum output current can be calculated using Equation (1), and the phase shift angle is equal to the case of the LCL network case because both networks are designed for the same maximum current. In the case of the LCL network, designing the phase difference between the inverter voltage and current is difficult because the values of L_r, C_r, and L_{lkg} are already fixed in order to satisfy the maximum output current. On the other hand, in the case of the LCCL network, the phase difference can be designed through the added compensation capacitor (C_{comp}), as shown in Equation (6). Figure 7b shows the phase difference of the LCCL network according to the load variation at the resonant frequency. As shown in Figure 7b, the LCCL network can ensure a wider phase difference than the LCL network. Therefore, the LCCL network can be operated in a wider ZVS range than the LCL network. In the case of the LCCL network, the inverter current can be calculated using Equation (7), and the inverter current of the LCCL network is increased compared with the inverter current of the LCL network.

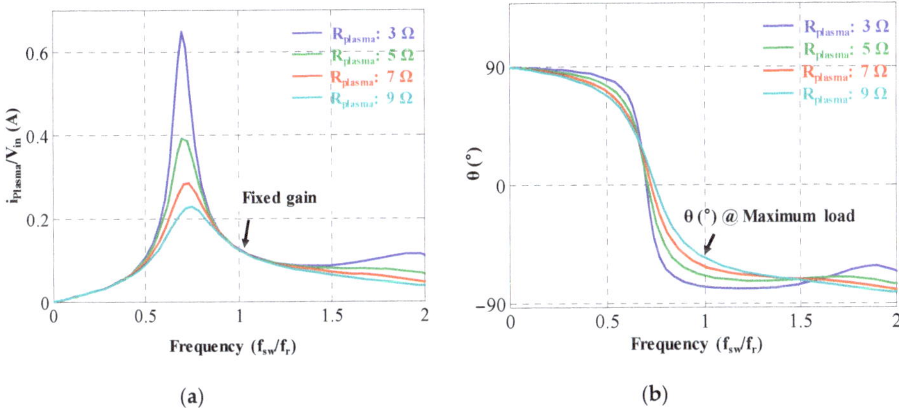

Figure 7. Characteristics of the LCCL network for designing a PCS: (**a**) output current characteristics of the LCCL network and (**b**) phase angle characteristics of the LCCL network.

Because of the compensation capacitor (C_{comp}), as mentioned previously, the inverter current of the LCCL network is higher than that of the LCL network, and the LCCL network can extend the ZVS region due to the compensation capacitor. Therefore, considering the overall efficiency of the inverter, loss analysis is necessary to compare the efficiency of both networks.

$$i_{plasma.LCCL}(f_r) = \frac{V_{o.inv.1}\sqrt{C_r}}{\sqrt{L_r}} \tag{5}$$

$$\theta_{LCCL}(f_r) = \tan^{-1}\left[\frac{\sqrt{L_rC_r}}{R_{plasma}C_{comp}} + \frac{(L_r - L_{lkg})}{R_{plasma}\sqrt{L_rC_r}}\right] \tag{6}$$

$$i_{o.inv.LCCL}(f_r) = \sqrt{\frac{C_r}{L_r}}\frac{1}{C_{comp}L_r\sqrt{\dfrac{1}{C_rL_r\left[C_rL_r+C_{comp}^2R_{plasma}^2+2C_{comp}(L_r-L_{lkg})\right]+C_{comp}^2(L_r-L_{lkg})^2}}} \tag{7}$$

4. Simulation and Experimental Results

The simulation and experimental results are presented for the purpose of verifying the analysis of the resonant network. The simulation has been conducted to describe the operation characteristics regarding the requirement and the phase shift angle. Finally, the applicability of the LCCL resonant network for the plasma generator is validated using experimental results. Figure 8 shows the simulation waveform about the LCL network and the LCCL network in the maximum current output condition. Both of the resonant networks can be operated to satisfy the maximum current output as shown in Figure 8a,b. When the LCL network is operated under the minimum current output conditions, the hard-switching region appears as is shown in Figure 8c. On the other hand, as shown in Figure 8d, the LCCL network can be operated with ZVS under the minimum load condition.

Loss analysis has been conducted to compare the efficiency of each resonant network [17]. In order to analyze the losses of the inverter, only the switch loss is considered, without any loss of resonant inductor, which consists of an air core inductor and a resonant capacitor. Regarding the analysis results, the loss of the LCL network is smaller than that of the LCCL network in the ZVS operation region under the 20 A output condition, as shown in Figure 9. In the non-ZVS operation region at the 20 A output condition, the loss of the LCL network is larger than that of the LCCL network. Beyond the 25 A output region, the loss of the LCCL network is larger than that of the LCL network because the large inverter current is required in the LCCL network.

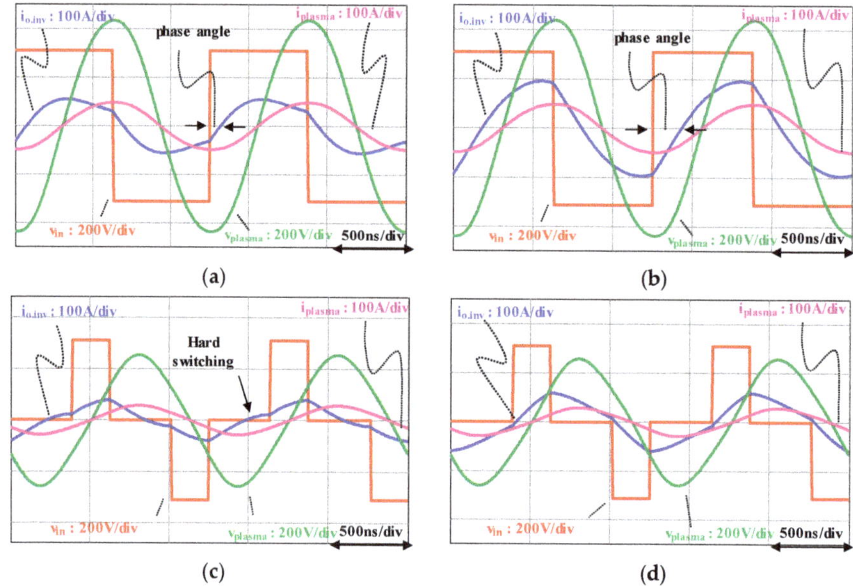

Figure 8. Simulation waveform of each resonant network: (**a**) characteristics of the LCL network under the 9 Ω, 35 A output conditions; (**b**) characteristics of the LCCL network under the 9 Ω, 35 A output conditions; (**c**) characteristics of the LCL network under the 9 Ω, 20 A output conditions; and (**d**) characteristics of the LCCL network under the 9 Ω, 20 A output conditions.

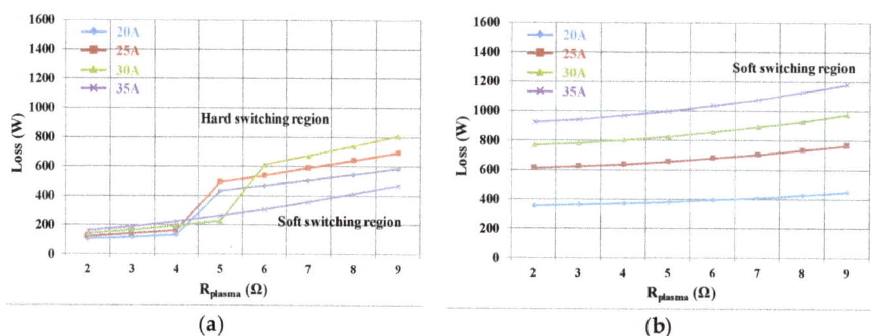

Figure 9. Loss analysis of each resonant network: (**a**) loss according to load variation in the LCL network; (**b**) loss according to load variation in the LCCL network.

The experiment has been conducted based on the parameters listed in Table 1. Among these parameters, the input voltage 1/3 scale-down model has been used with consideration of the laboratory power distribution. The resonant network has been designed with Equation (2) considering switching frequency and maximum output current. Figure 10a shows the experimental waveform of the LCL network, which represents the narrow phase angle. In this case, it is impossible for i_{plasma} to be controlled with the ZVS region using the phase shift control to obtain a minimum value, as shown in Figure 10a. On the other hand, a sufficient phase angle for the phase shift control in the ZVS region is obtained in the LCCL network, as shown in Figure 10b. In this result, the DC-AC inverter adopting the LCCL network can be operated with ZVS despite the worst case for obtaining the phase angle, as shown in Figure 10c.

Table 1. Parameters of the inverter and both resonant networks for simulation and experiment.

Parameters	Value (Unit)	Parameters	Value (Unit)
Input voltage (V_{in})	311 (V_{dc})	Leakage inductance (L_{lkg})	2 (uH)
Switching frequency (f_{sw})	400 (kHz)	Equivalent resistor of plasma load (R_{plasma})	2–9 (Ω)
Plasma current (i_{plasma})	20–35 (A_{rms})	Compensation capacitance (C_{comp})	39.6 (nF)

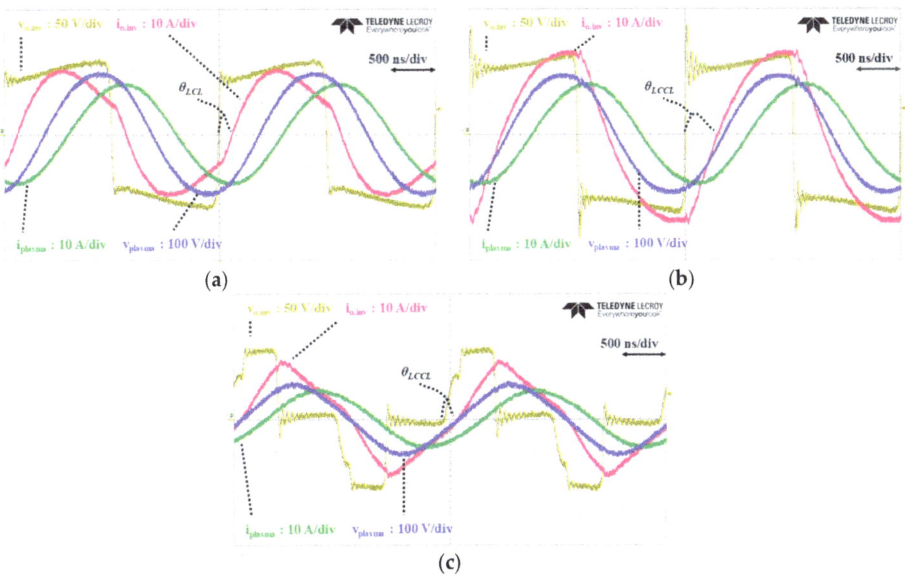

Figure 10. Feasibility test results of experimental waveform of each resonant network: (**a**) experimental waveform of the LCL network under the 9 Ω, 11.6 A conditions; (**b**) experimental waveform of the LCCL network under the 9 Ω, 11.6 A conditions; and (**c**) experimental waveform of the LCCL network under the 9 Ω, 6.6 A conditions.

The phase difference and soft-switching ability of the designed networks are verified through these experimental results. In the resonant inverter for using the plasma generator, the inverter loss is determined with the trade-off relationship between the conduction loss of the inverter current and the switching loss due to the phase difference. Therefore, the LCCL resonant network could be better performing than the LCL network under the large resistance of the plasma load and small plasma current conditions.

5. Conclusions

In this paper, the applicability of the LCL network and LCCL network for a plasma generator have been investigated with consideration of the characteristics of the plasma load. In particular, the drop-out phenomenon, which is substantially different from the resistive load, is considered for the design of an LCL network and an LCCL network. Considering the characteristics of the plasma load, an LCL network and an LCCL network are designed and analyzed. Based on the analysis, a simulation was conducted in order to verify the validity of the designed networks. Furthermore, the experiment was progressed while considering the laboratory power distribution level in order to verify the applicability of the networks.

Author Contributions: Conceptualization, G.W.K. and W.-Y.S.; methodology, G.W.K.; software, G.W.K. and W.-Y.S.; validation, G.W.K.; formal analysis, G.W.K.; investigation, G.W.K.; resources, B.K.L.; data curation, G.W.K.; writing—original draft preparation, G.W.K. and B.K.L.; writing—review and editing, G.W.K. and B.K.L.; visualization, G.W.K. and W.-Y.S.; supervision, B.K.L.; project administration, B.K.L.; funding acquisition, B.K.L.

Energies **2019**, *12*, 3156

Funding: This work was supported by "Human Resources Program in Energy Technology" of the Korea Institute of Energy Technology Evaluation and Planning (KETEP), granted financial resource from the Ministry of Trade, Industry & Energy, Korea. (No. 20184030202190); This work was supported by the Korea Institute of Energy.

Conflicts of Interest: The authors declare no conflict of interest.

References

1. Shimizu, K.; Kinoshita, K.; Yanagihara, K.; Rajanikanth, B.S.; Katsura, S.; Mizuno, A. Pulsed-Plasma Treatment of Polluted Gas Using Wet-/Low-Temperature Corona Reactors. *IEEE Trans. Ind. Appl.* **1997**, *33*, 1373–1380. [CrossRef]
2. Kim, D.W.; You, S.J.; Kim, J.H.; Chang, H.Y.; Oh, W.Y. Computational Characterization of a New Inductively Coupled Plasma Source for Application to Narrow Gap Plasma Processes. *IEEE Trans. Plasma Sci.* **2015**, *43*, 3876–3882. [CrossRef]
3. Millner, A.R. Power Electronics Topologies for Plasma Generators. In Proceedings of the 2008 IEEE International Symposium on Industrial Electronics, Cambridge, UK, 30 June–2 July 2008; pp. 359–362.
4. Tran, K.; Millner, A. A New Power Supply to Ignite and Sustain Plasma in a Reactive Gas Generator. In Proceedings of the 2008 Twenty-Third Annual IEEE Applied Power Electronics Conference and Exposition, Austin, TX, USA, 24–28 February 2008; pp. 1885–1892.
5. Ahn, H.M.; Sung, W.-Y.; Lee, B.K. Analysis and Design of Resonant Inverter for Reactive Gas Generator Considering Characteristics of Plasma Load. *J. Electro. Eng. Technol.* **2018**, *13*, 345–351.
6. Chabert, P.; Braithwaite, N. Impedance of the Plasma Alone. In *Handbook of Semiconductor Wafer Cleaning Technology*; Cambridge University Press: Cambridge, UK, 2011; pp. 233–235.
7. Lee, S.Y.; Gho, J.-S.; Kang, B.-H.; Cho, J.-S. Analysis of Pulse Power Converter for Plasma Application. In Proceedings of the 2008 IEEE International Symposium on Industrial Electronics, Cambridge, UK, 30 June–2 July 2008; pp. 556–560.
8. Kuroki, T.; Mine, J.; Okubo, M.; Yamamoto, T.; Saeki, N. CF_4 Decomposition Using Inductively Coupled Plasma: Effect of Power Frequency. *IEEE Trans. Ind. Appl.* **2005**, *41*, 215–220. [CrossRef]
9. Kuroki, T.; Mine, J.; Odahara, S.; Okubo, M.; Yamamoto, T.; Saeki, N. CF_4 Decomposition of Flue Gas from Semiconductor Process Using Inductively Coupled Plasma. *IEEE Trans. Ind. Appl.* **2005**, *41*, 221–228. [CrossRef]
10. Rooms, S.D. A Plasma Torch Converter Based on the Partial Series Resonant Converter. In Proceedings of the AFRICON Conference, Windboek, South Africa, 26–28 September 2007; pp. 1–6.
11. Chen, G.; Sun, Q.; Hu, T.; Guo, Q. A Novel Digital Control Algorithm for a DC-DC Converter in Plasma Application. In Proceedings of the 2011 Asia-Pacific Power and Energy Engineering Conference, Wuhan, China, 25–28 March 2011.
12. Pacheco-Sotelo, J.O.; Barrientos, R.V.; Pacheco-Pacheco, M.; Flores, J.F.R.; García, M.A.D.; Benitez-Read, J.S.; Peña-Eguiluz, R.; López-Callejas, R. A Universal Resonant Converter for Equilibrium and Nonequilibrium Plasma Discharges. *IEEE Trans. Plasma Sic.* **2004**, *32*, 2105–2112. [CrossRef]
13. Pacheco-Sotelo, J.; Peña-Eguiluz, R.; Eguiluz, L.P.; Ríos, A.S.; Sanchez, G.C. Plasma Torch Ignition by a Half Bridge Resonant Converter. *IEEE Trans. Plasma Sic.* **1999**, *27*, 1124–1130. [CrossRef]
14. Amjad, M.; Salam, Z.; Facta, M.; Mekhilef, S. Analysis and Implementation of Transformerless LCL Resonant Power Supply for Ozone Generation. *IEEE Trans. Power Electron.* **2013**, *28*, 650–660. [CrossRef]
15. Ahn, H.M.; Jang, E.S.; Ryu, S.; Lim, C.S.; Lee, B.K. Control Strategy for Power Conversion Systems in Plasma Generators with High Power Quality and Efficiency Considering Entire Load Conditions. *Energies* **2019**, *12*, 1723. [CrossRef]
16. Borage, M.; Tiwari, S.; Kotaiah, S. Analysis and Design of an LCL-T Resonant Converter as a Constant-Current Power Supply. *IEEE Trans. Ind. Electron.* **2005**, *52*, 1547–1554. [CrossRef]
17. Graovac, D.; Purschel, M.; Kieo, A. *MOSFET Power Losses Calculation Using the Data-Sheet Parameters*; Technical Report for Infineon: Dresden, Germany, July 2006.

MDPI
St. Alban-Anlage 66
4052 Basel
Switzerland
Tel. +41 61 683 77 34
Fax +41 61 302 89 18
www.mdpi.com

Energies Editorial Office
E-mail: energies@mdpi.com
www.mdpi.com/journal/energies